ONLY JUST

A HISTORY OF
THE A34 NEWBURY BYPASS
1979 - 1998

Gordon   Rollinson

Published by
Gordon Rollinson
3 Lovett Walk
WINCHESTER
SO22 6NL

ISBN No 0 9531861 0 5

ACKNOWLEDGEMENTS

I gratefully acknowledge the comments, suggestions, help or encouragement I have received in the preparation of this work from Sir David Mitchell, David Rendel M.P., Peter Davies, Jack Smith, colleagues in the Newbury Bypass Supporters' Association and other Newbury friends too numerous to name. I also acknowledge the use I have made of reports in past columns of the Newbury Weekly News for quotations and as an aide memoire and the permission of its Group Editor to reproduce headlines from copies of the NWN and The Advertiser. I thank the London Evening Standard for permission to reproduce the JAK cartoon and the Ordnance Survey for permission to reproduce the area plan of the bypass routes in the centre pages and on the cover. The photographs referenced ©PB are reproduced with the kind permission of Peter Bloodworth of Newbury.

DISCLAIMER

The contents of this book are correct to the best of my knowledge and belief but no responsibility is accepted for errors and omissions.

GR, Nov 1997

CONTENTS

INTRODUCTION

She burst through the door. She was well-dressed and in her late thirties, I guessed. "How can you encourage the destruction of so much countryside?", she said furiously, "How can you support such a scheme ?".

It was March 1995. I was helping with an exhibition in Newbury's main shopping centre in support of the bypass. The centrepiece was a three-metre model of the proposed road whose construction had been postponed to some indefinite future date - "put on hold" in the jargon of the Civil Service - by the Secretary of State for Transport, Dr Mawhinney, three months previously. The exhibition was acting as the focal point for those local people (a 6:1 majority according to media polls) who wanted the western bypass built as quickly as possible and wished to sign the petition, organised by the area's two M.P.s, to bring home to Dr Mawhinney their anger and frustration at the delay.

It was inevitable that such an event would also attract the attention of both the anti-road protesters and a few local residents, such as the angry woman who now confronted me, who had either not troubled, or perhaps had not had the opportunity, to keep up with the progress of the bypass inquiries and procedures. I guessed that her knowledge of the issues was limited to absorbing the protesters' disinformation, recently fed to those media news hounds who scented the newsworthiness of another Twyford Down.

Attempting to pacify her I said, "Can we at least agree on this? Newbury's awful traffic problems need to be resolved. Let the Government appoint consultants to carry out an in-depth investigation of the reasons for the congestion and produce and publish the possible solutions. Then allow any member of the public to discuss those solutions with the consultants and with Department of Transport's officials. After carefully considering the public's responses, let the DoT publish its chosen solution and give everybody concerned - landowners, residents, motoring organisations, anti-roads bodies, environmental agencies - a sufficiently long period of time, say a year or two, to carry out their own investigations and develop their own alternatives. After that let us have an open-ended public inquiry, giving higher priority to landscape and ecology matters than cost. Let any interested party, organisations or individuals come along to promote and defend their schemes."

She had calmed down. "Now, that is sensible," she said, "That's what should be done."

"We're not finished yet," I said, "Even when the optimum solution has been found, there could still be householders, farmers or other local people who might be affected by compulsory purchase orders or by the proximity of access roads, embankments, tree planting or other details. So let's have another intensive public inquiry to ensure that their rights are not being trampled on."

"That's very fair", she said, "So why don't they do it ?"

"But," I said patiently, "that is exactly what has been happening over the past sixteen years." She stepped back. "I don't believe you," she said furiously, slamming the door as she left.

She is not alone in her attitude. The protest movement against the Newbury Bypass has been characterised by ignorance of its history, coupled with an unwillingness to be enlightened. The long-suffering people of Newbury see the bypass as far too long delayed by dilatory government processes and far too expensive due to the actions of militant anarchists. The committed road protesters will brush aside arguments which support the building of any road, anywhere. They proclaim that the application of measures such as better public transport, cycle ways and parking restrictions will eliminate congested roads. Those more intimately involved with the assessment of a particular scheme come to realise that the complexity of the issues do not allow the luxury of such a simple conviction. Other reasons for building a bypass come into the equation. The protecting of human life must be of prime importance. Then there is the overall improvement in the quality of living for the majority of those living and working in the town by the reduction of noise, atmospheric pollution and visual intrusion. Ecological conservation is high on the list; alternative schemes may cause even more damage to the environment. Also, though of lesser importance, into the balance must go the considerable cost-savings to individuals and to industry which a bypass can bring.

1

Newbury is a typical example of a town needing a bypass. While the details may differ significantly, the general case for a Newbury bypass also applies to bypasses elsewhere and the processes used at Newbury are the standard procedures adopted for assessing public works. This summary of the issues at Newbury - and it can be no more than that - is written from the perspective of one who has lived in the area for thirty-five years and, though a country-lover, walker, cyclist and regular public transport user, has seen the justification for a western bypass at Newbury become overwhelming.

GR 1997

Traffic on the A34 bisects Newbury. The regularly-choked Sandleford Link into Newbury town centre carries a high-than-average proportion of HGVs due to the south coast docks traffic

CHAPTER 1

Location and History

Newbury is a historic and relatively prosperous small town, 55 miles from London, situated in the centre of southern England roughly at the intersection of an east/west line from London to Bristol and a north/south line from Oxford to Southampton. Although set in the countryside, it is the base for numerous small and medium sized companies who provide the area with higher than average employment. The town is also home to commuters who work in London, Reading, the atomic energy establishments at Harwell and Aldermaston, and elsewhere. Administratively, Newbury District Council encompasses a population of approximately 140,000, about half of whom live in the built-up areas in and around Newbury, including Thatcham to the east where the majority of the housing and commercial development has taken place in recent years. The other half live in the surrounding countryside and villages.

The centre of Newbury lies in the valley of the River Kennet, a tributary of the Thames, and is about 75 metres above sea level. To the north and south, higher ground provides sites for various commons, the most widely known being Greenham Common. (Its abandonment as an airfield in 1993 opened up the possibility of another bypass route.) The valley consists of gravel beds with London clay and peat overlaying chalk. The commons generally consist of woods and heathland. Newbury Racecourse, opened in 1905, occupies over ½ square mile of open ground between Newbury and Thatcham.

There is evidence of inhabitants in the area before 2,500 BC, Mesolithic implements having been discovered near the centre and to the east of the town. Neolithic trackways ran along the Berkshire Downs to the north and along the North Hampshire Downs to the south, including Watership Down. The large earthworks associated with Iron Age forts can be seen on several of the surrounding hills. Roman relics and traces of Roman roads have been discovered in and around the town. (Archaeological investigations would clearly be required for any bypass route.) The name Newbury is first mentioned in 1079 and it is referred to as a borough in 1189. There is now no sign of the central castle, captured by Stephen in 1153, but 1½ miles NW of the town centre, Donnington Castle, rebuilt in 1386 and ruined by Cromwell after the Civil War, still remains. The town became a major centre for the wool trade in the 15th and 16th centuries and its leading citizen, Jack of Newbury, became sufficiently wealthy to entertain Henry VIII and his court.

The best known local historic events were the two major Civil War battles, whose sites were to be of significance in the bypass investigations. The first of these took place on 20 September 1643 to the SE of the town in the areas of Skinners Green, which still consists of open fields, and Wash Common, which is largely covered by housing development. The second battle, on 27 October 1644, took place over an area between Speen and Donnington to the north-west of the town centre. All of this site is now built upon. There was a further skirmish within the town on 9 November 1644, recorded in the Newbury Museum as the Third Battle of Newbury.

Traffic Within Newbury

Since the closure of the Winchester to Didcot railway line, the sole north/south route through Newbury has been the A34 trunk road connecting Winchester to Oxford. There are three east/west routes which all follow the valley of the River Kennet. They consist of the A4 trunk road (substantially the old coaching road joining London to Bath), the Reading to Westbury section of Railtrack's western railway line and the Kennet and Avon canal. Together with the river itself, they provide impediments to the construction of any new north/south road.

It may seem almost beyond belief to those familiar with Newbury's present problems that, until 1966, all north/south trunk road traffic had to negotiate the main shopping street, Northbrook Street, (see Newbury street plan on page 4) passing over the combined river and canal by the attractive eighteenth-century stone arch Town Bridge, now restricted by traffic lights to single lane alternate flows. The only other river/canal crossing within the town was the inappropriately named "American"

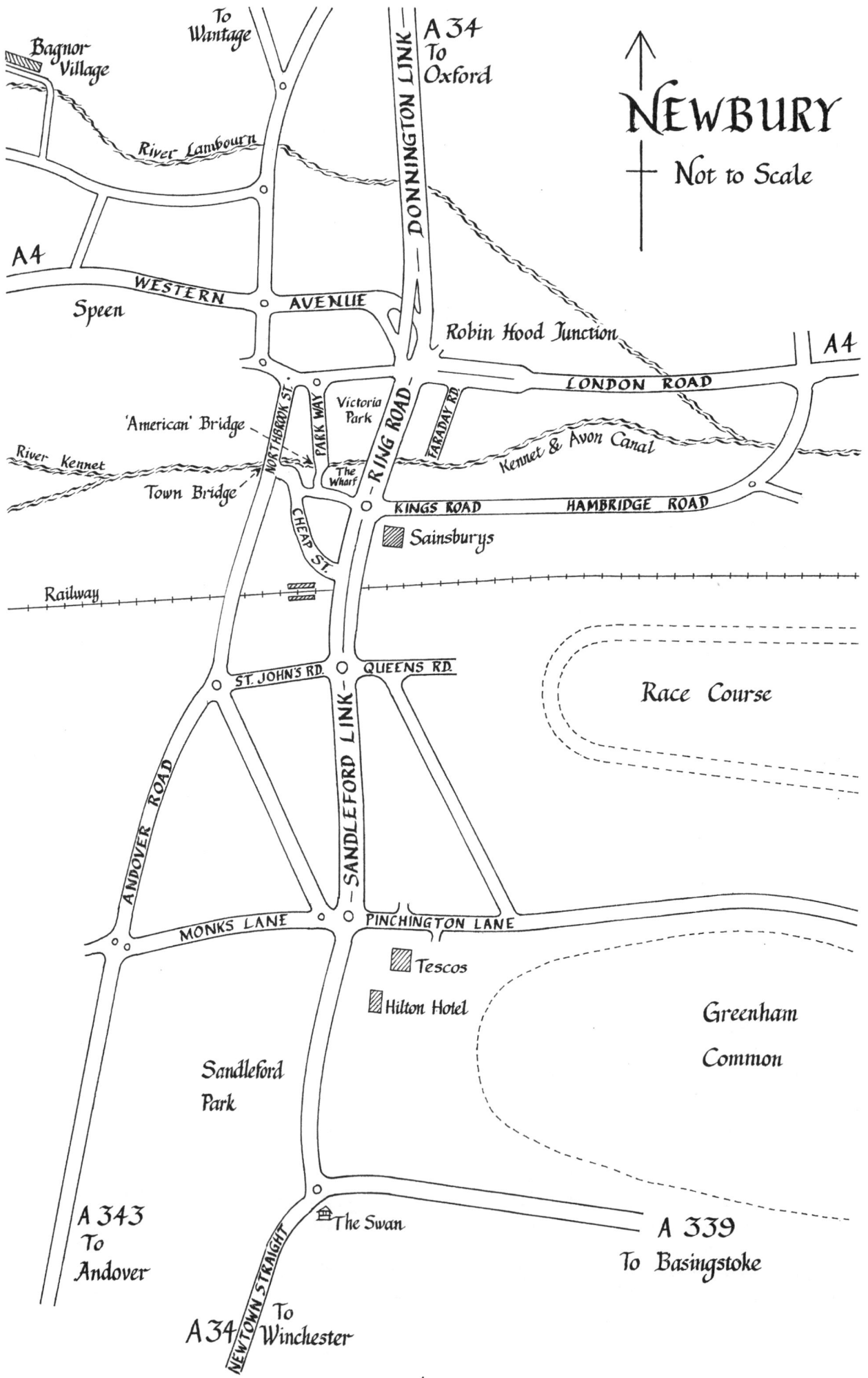

NEWBURY

Not to Scale

Bagnor Village

To Wantage

A34 To Oxford

DONNINGTON LINK

River Lambourn

A4

WESTERN AVENUE

Speen

Robin Hood Junction

LONDON ROAD

A4

NORTHBROOK ST.

PARK WAY

Victoria Park

RING ROAD

FARADAY RD.

'American' Bridge

River Kennet

Kennet & Avon Canal

Town Bridge

The Wharf

KINGS ROAD

HAMBRIDGE ROAD

CHEAP ST.

Sainsburys

Railway

ST. JOHN'S RD.

QUEENS RD.

Race Course

SANDLEFORD LINK

ANDOVER ROAD

MONKS LANE

PINCHINGTON LANE

Tescos

Hilton Hotel

Greenham Common

Sandleford Park

A343 To Andover

NEWTOWN STRAIGHT

The Swan

A339 To Basingstoke

A34 To Winchester

Bridge, a one-way single-lane steel girder bridge, constructed in 1940, connecting Park Way to the ancient Wharf area. East/west traffic between London and Bristol had to traverse the A4 through Newbury .

Even in the late 1950s, Newbury's roads were unable to cope with the traffic volumes put upon them. An east/west relief road, the Western Avenue, was opened in 1959 and connects the Robin Hood junction to Speen. In 1961, the Ministry of Transport appointed a Study Group under the leadership of Colin Buchanan to *"study the long term development of roads and traffic in urban areas and their influence on the urban environment"*. Their report, "Traffic in Towns", was published in 1963 and considered in depth four practical examples. Newbury was chosen as the small town model, and some 25 pages of the report are devoted to a study of the town's traffic routes and proposals, with costs, for improving them.

Buchanan assumed that both east/west and north/south bypasses would be provided at Newbury. The M4 South Wales motorway, which would become the east/west bypass, was then only in the planning stages. A diagram in his report indicates a western route for the north/south bypass, although no formal proposals had been published at the time.

To cope with the growth of local traffic, Buchanan went on to describe a possible primary network of roads within Newbury which he considered should be built first, indicating that when they became congested, the outer bypasses should then be built. Even while the Buchanan report was being written, a dual-carriageway improvement to the A34 north/south route within the town was being designed. It was not to follow Park Way on the western side of Victoria Park, as Buchanan had suggested, but instead kept to the eastern side. It provided new north/south crossings of the river/canal and the railway but its title of "Ring Road" is somewhat misleading. It is more appropriately called the Inner Relief Road. It was eventually opened in 1966 after constructional delays to the bridges. However, the sweeping improvements which Buchanan proposed for London Road and Hambridge Road have not been built, although some major improvements, particularly to junctions, have been carried out.

The Maidenhead to Liddington section of the east/west M4 motorway to the north of Newbury was opened in 1971. To provide an adequate connection from the new M4 to Newbury, the so-called Donnington Link was built, being a replacement of the A34 between Junction 13 of the M4 at Chieveley and the Robin Hood roundabout at the northern end of the Newbury Ring Road. The east/west through traffic in Newbury was immediately relieved and traffic volumes remain comparatively manageable on the A4 to the west of Newbury and on the A4 to the east of Thatcham.

However, the rapid increase in local traffic over the past 25 years has caused major problems of congestion since a high percentage of local journeys cross the busy A34. Local and through traffic intermingle along a ½ mile section of the A34 within the central area of Newbury causing queues which can extend for several miles to the north and south of the town at the weekend peaks. The problem has been exacerbated by the higher-than-average increase in the through north/south traffic on the A34. This is perhaps best examined by considering the flow along the two-mile single carriageway section of the A34 to the south of Newbury which has become known (to the irritation of many living there) as the Newtown Straight. This stretch carries the highest percentage of through traffic - approximately 75% - and therefore the influence of local traffic is less pronounced. A traffic counter installed at Tot Hill to the south of the Newtown Straight showed an increase from 11,000 AADT (Annual Average Daily Traffic flows) in 1981 to 26,500 in 1995[1], an increase of 140%. (See chart on page 45.) Over the same period, general traffic in the UK increased from 276.9 billion vehicle kilometres to 421.9[2], an increase, by comparison, of only 52%.

The reasons for this escalation in traffic were due mainly to the improvements to the A34, both within Newbury and also to the north and south, which have made it the principal direct north/south route through central southern England. (See roads diagram on page 8.) The ¾mile Sandleford Link leading out of Newbury town centre to the south was opened in 1977. The 2 mile Whitway Diversion, four miles south of Newbury, was given approval in August 1986 and was opened early in 1989. Also in 1989, an extensive ten-traffic-light junction was built at the Robin Hood roundabout in Newbury. In

1990 came the extension of the M40 motorway between Birmingham and Oxford with a much improved access to the A34 immediately to the north of Oxford.

The general A34 improvements not only caused a increase in traffic volume, but also produced a dangerous mix of vehicles at Newbury. Transport to and from the docks provides a high percentage of HGV traffic on the A34. For example, around 20% of total traffic at Tot Hill consists of HGVs compared with the average UK trunk road figure of 12%[3].

The Rail Alternative

In common with other bypass schemes, it has been argued for some years by opponents that more investment in the rail network would reduce, if not eliminate, the need for a north/south bypass at Newbury. The north/south railway line, whose 1885 Newbury/Winchester section finally closed in 1961, was single-track only. A superior twin track north/south connection avoiding Newbury is provided by the railway line through Oxford, Reading and Basingstoke which already carries car-transporter and container trains from the docks. Cross-country passenger trains run daily on this route between Scotland, the North and the Midlands to Southampton and the Bournemouth area. The existing railway at Newbury is already well used by commuters travelling in an easterly direction to and from London and Reading.

Railheads, for the transfer of freight between road to rail, cannot be sited at sufficiently close intervals to be economic for most users when compared with the comprehensive road network. The cost of railhead transfer and the comparatively short haul distances within the UK have not made rail freight sufficiently attractive to prevent road freight increasing fivefold in the past forty years with rail freight decreasing by one half in the same period. The changes in industrial and retail practices have also had a considerable effect. "Just in time" delivery techniques reduce factory storage costs or enable the freshest produce to be offered by retailers to their customers. Faster delivery of goods and spare parts are achieved by vans continuously shuttling between depots and dealers. The operator can offer a more efficient service to his customers by having the departure and arrival time of his goods under his own control and not limited to a railway timetable.

Some special cases of bulk material transport do not require such precise timing. For example, at Newbury railway station, freight trains carrying crushed stone from west country quarries to the London area are a familiar sight; bulk movements which otherwise would cause some extra 2000 heavy vehicle movements per day on the M4 motorway. Few opportunities exist for similar major uses of rail transport which would remove traffic from the roads. "Piggy-back" rail wagons to carry HGVs within the UK, advocated by some anti-road organisations and used on some continental lines, would not accommodate the largest standard road trucks on most existing UK railway lines without the enormously expensive and lengthy task of raising bridges, together with their approach roads.

However some growth in rail freight can be expected on long-haul and international routes. During 1997, the total volume of rail freight on British railways actually showed an increase, for the first time in 50 years. The growth was accounted for by long-distance freight, but it was only a modest 3%. Given Channel Tunnel access to continental high-speed lines and Railtrack's substantial investment proposals for British main lines, the trend is likely to continue, but it would be quite unrealistic to expect it to have influence on whether or not to build a scheme such as the Newbury Bypass.

Prevention of Traffic Growth

Most road users in this country have come to recognise that the inexorable growth of road traffic over the past two or three decades must eventually be curtailed. This will not be achieved simply by putting an embargo on all road building, but by many other measures. Parking will become even more restricted as will vehicle access to some city streets. Park and ride schemes are proving to be effective in relieving some traffic from town centres and are being extended where feasible. The RAC believes that commuting traffic will be reduced by a wider use of computers at home. But the principal disincentives will need to be financial: ever-increasing road and fuel taxes, road tolls and parking charges. Lifestyles have become so dependent on private transport that, for such financial impositions to be effective in weaning the public away from the motor car, they will need to be set at a very high level. There is attraction to any Government in these additional taxes, which would bring in enormous revenues. In 1997 the overall contribution in taxes to the Exchequer by road users amounted to some £28 billion of which much less than half was spent on roads and public transport subsidies.

However, increasing taxes by the necessary amounts would appreciably increase inflation and affect the economy to an extent which means that, politically, they could only be introduced over a very long time-scale. Such extra taxes would also be regarded as unjust in penalising that section of the community least able to afford them. They would also unfairly affect country dwellers who have to travel the greatest distances and are least likely to be served with a public transport alternative. Successive governments have found it difficult to transmute their expressed support for an integrated transport policy into concrete actions.

While the majority of car users acknowledge that congested roads can only be relieved by lower car usage, it is not being too cynical to say that most believe that it is others who could use their cars less frequently, they themselves having particular reasons for not doing so. The sheer convenience and privacy of their own door-to-door transport is compelling. A new generation has grown up wedded to car transport. For those unused to using buses and trains, the change would be regarded as a reduction in their standard of living which they would accept only if the financial savings were much greater than those at present. Older generations would regard going back to using public transport as a retrograde step. Even the most comprehensive public transport system would provide only a tiny fraction of the population with a service outside their door. The remainder would need to prepare for a possibly cold, wet and windy walk and a wait at their nearest bus stop, only to join a swaying mode of transport, particularly unpleasant for the elderly and infirm or mothers with pushchairs, in company with passengers not of their choosing. Many parents would flatly refuse to send their children to school by public transport, still less by bicycle, until, rightly or wrongly, they perceive a large reduction in the danger of criminal actions in society. The established shopping pattern of bulk supermarket purchases by car at weekly or greater intervals could not readily be superseded by public transport.

A re-education of the car user is required. When comparing the expense of public transport with a similar journey by car, motorists tend to take into account only the immediate costs of fuel and parking. Other running costs, such as servicing and wear & tear, are ignored. Seldom entering into the equation are a proportion of the fixed costs such as insurance, road fund tax, MoT testing or, greatest of all, depreciation. Few even bother to car-share. To those who now cycle or regularly use public transport, it is frustrating to accept, but nevertheless has to be recognised, that the necessary changes in habits and lifestyles will probably be accomplished over decades rather than years.

There could be no question of providing an alternative to a bypass at Newbury by the introduction of the long term measures which will eventually be needed to curtail traffic growth nationally. The time scale to save the town of Newbury had to be much shorter.

The Newbury Strategic Bottleneck (see map on page 8)
Within the United Kingdom, the A34 passing through Newbury forms the principal connection between the industrial Midlands and North to the south coast, and in particular to the docks at Southampton, Portsmouth and Poole. Internationally, the section of the A34 at Newbury is part of Euroroute E05 (though not signposted as such in the UK), which extends for over 1,600 miles (2,550 kms) from Greenock on the Firth of Clyde via Southampton, Le Havre, Paris and Madrid to Algeciras in southern Spain, across the bay from Gibraltar. This route has been improved to 4-lane dual carriageway construction along most of its length. The narrowest section is the 3-mile section south of Newbury which includes the 7-metre-wide 2-lane single carriageway Newtown Straight and which, predictably, has a rising accident rate, contrary to the national trend. Approaching Newbury from the south, five-mile queues of traffic between Beacon Hill and the town centre have become commonplace at peak periods. The situation on the north side of the town is little better. It is quite common to take ¾ hour to cover the ten mile section of the A34 through the Newbury area.

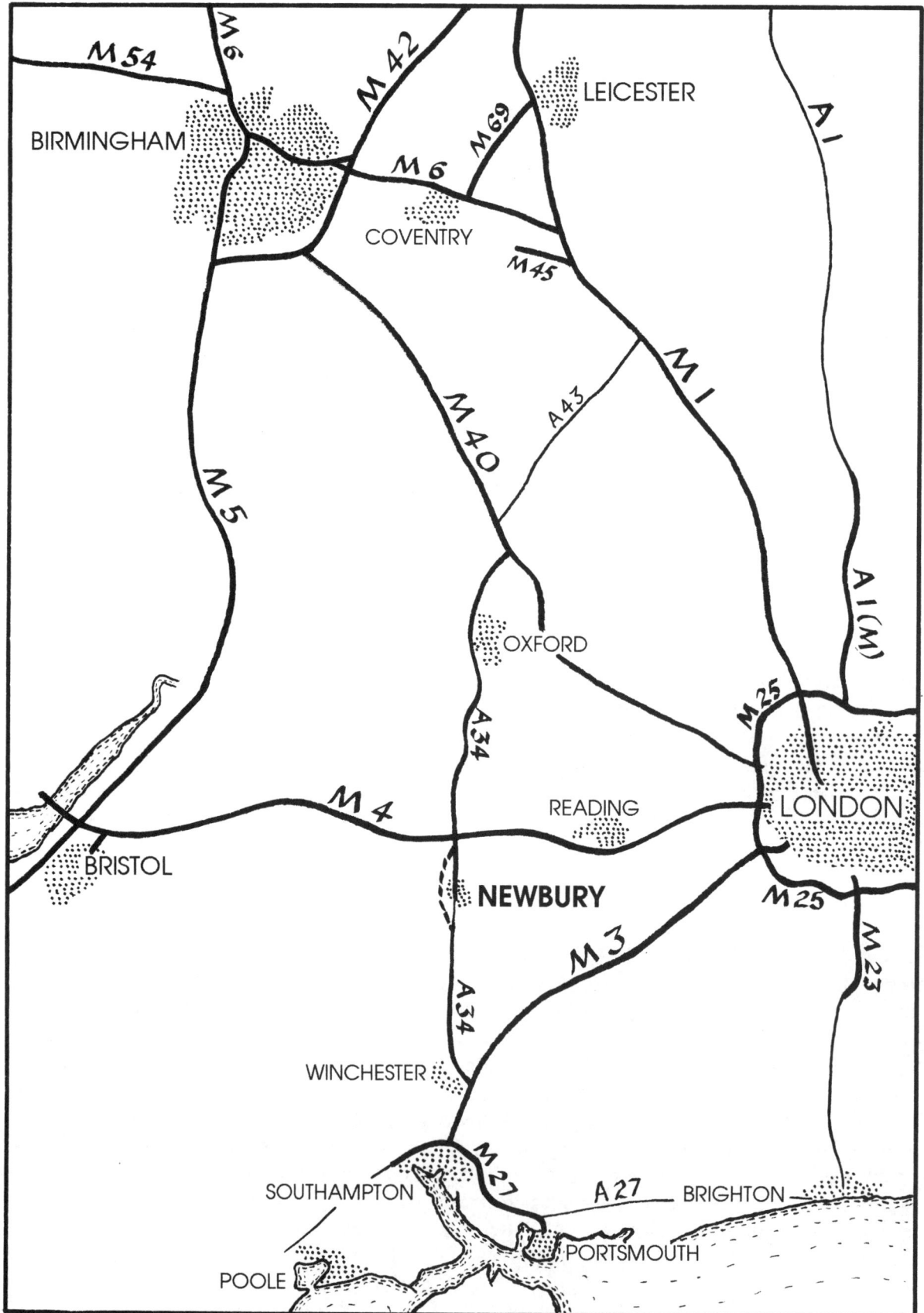

POSITION OF NEWBURY IN THE
MOTORWAY AND TRUNK ROAD NETWORK

M 54
M 6
M 42
M 69
LEICESTER
A 1
BIRMINGHAM
M 6
COVENTRY
M 45
M 1
A 43
M 40
M 5
A 34
OXFORD
A 1(M)
M 25
M 4
READING
LONDON
BRISTOL
NEWBURY
M 25
A 34
M 3
M 23
WINCHESTER
M 27
SOUTHAMPTON
A 27
BRIGHTON
PORTSMOUTH
POOLE

In Newbury itself, the combined effect of the increases in local and through traffic confirmed undeniably Buchanan's 1963 portentous summary, *"The penetration of motor vehicles throughout urban areas is bringing its own peculiar penalties of accidents, anxiety, intimidation by large or fast vehicles that are out of scale with the surroundings, noise, fumes, vibration, dirt and visual intrusion on a vast scale".*

It was clear that the two issues of local and through traffic had to be addressed, but Buchanan's recommendation of "local solutions first" could not be implemented. The Local Authorities concerned, Berkshire County Council and Newbury District Council, were only too aware of the need for improvements to local roads but a proper assessment of local transport needs, let alone the funding, design and construction of relief measures such as those proposed by Buchanan, could not be put in hand until the question of a north/south bypass had been resolved. The bypass route, the location of its junctions and its opening date would all have considerable impact on the decisions which had to be made for the approval and financing of local road improvements and for the granting of planning permission for other developments. The "bypass" could well have been routed through the middle of the town. Settling the line of the bypass became an essential first step.

Although tentative routes had been mooted by the Department of Transport (DoT) in the 1960s, it was not until May 1979 that the DoT appointed their consultants, Mott Hay and Anderson, to begin work in earnest with a brief, *"to investigate the section of the A34 between Tot Hill and Donnington, to proceed to Public Consultation and to recommend a solution"* .

References:
[1] source, Highways Agency
[2] source, British Road Federation, Transport Statistics Great Britain
[3] source, Statement of Reasons, Departments of Transport & Environment, 1988

An accident on the single-carriageway Newtown Straight, the narrowest section of the A34 and of Euroroute E 05 which extends from Greenock, on the Clyde, to Algeciras in southern Spain

CHAPTER 2

Public Consultation

Mott Hay & Anderson, (now Mott McDonald) compiled two reports. The first, dated March 1980, was an Interim Report, identifying the route corridors to be studied and highlighting their relative merits. 37 route options were assessed and the consultant's principal comments related to the environmental problems likely to be encountered in the construction of each route. The second report was a Technical Appraisal recommending alternative routes to be put forward for public consultation.

There is no doubt that the DoT was being more circumspect over its relations with the public, doubtless influenced in large measure by the problems it had encountered at Winchester. In June 1976, the opening of the second public inquiry into the M3 route around Winchester had been disrupted by protesters in the glare of wide media coverage. The line of the M3 between Popham, south west of Basingstoke, to Compton, south of Winchester, had already been investigated at an earlier inquiry in 1971 and given approval by the Secretaries of State in 1973. The later, 1976 inquiry was convened to consider ensuing matters such as side road orders, connecting road schemes, compulsory purchase orders and exchange land certificates. However, by the actions of the protesters and by listening to other arguments, the Inspector at Winchester had been persuaded that the public interest would best be served by widening the scope of the inquiry to reconsider the actual line of the M3. Alternative routes were put forward and examined and as a consequence, the inquiry was not completed until twelve months later, in June 1977. It took another three years for the Inspector's report to be compiled, assessed and published and a decision made to initiate a new study into the M3 route. By 1980 it was clear that the road construction would be delayed by many years. (The final section of the road was not open until 15 December 1994, fourteen years later than the date forecast at the 1976 inquiry.) The consulting engineers, Mott Hay and Anderson, were given a brief by the DoT in 1981 to consider a new scheme for Winchester. The brief included the following significant statement, *"The new Scheme will need to have the highest possible measure of acceptance by the Local Authority, other interested parties and, most important of all, by the general public."*

With this guiding precept, it was clear that, at Newbury, the DoT would endeavour to avoid a repetition of the problems it had encountered at Winchester by ensuring that the public were consulted at an early stage of the design.

A34 Newbury

Tot Hill to Donnington

The Future Route ?

Department of Transport,
South East Regional Office,
Federated House,
London Road,
Dorking,
Surrey RH4 1SZ

**Please give us
your views**

In 1982, a brochure was prepared which explained that the traffic problems at Newbury could not be resolved by traffic management schemes or minor highway improvements alone and that more radical measures were needed. The brochure showed, to a scale of 1/25,000, four possible bypass routes; a western route (the brown route), two variants of a central route (the orange and purple routes) and an eastern route (indicated by a dashed line) which was not being recommended but which was included for comparison purposes. The western route was a logical choice as it used three miles of the defunct Newbury/Winchester railway line. The brochure also showed the location of Areas of Outstanding Natural Beauty, Areas of Great Landscape Value, MoD land, Common Land, Recreation Areas, Industrial Areas, Conservation Areas, Sites of Special Scientific Interest (SSSIs), Scheduled Ancient Monuments, Nature Reserves, Archaeological Sites and Listed Buildings.

It gave the estimated cost of each route, varying from £18.3M for the western route to £27.3M for the most expensive central variant. It also gave the amount of agricultural, woodland and other types of land which would be taken; the number of buildings which would need to be demolished; the number of dwellings which would be subject to an increase or a decrease in noise levels; the features of

10

each road design such as junctions, embankments and cuttings; the traffic advantages of each route and, most importantly, it highlighted the impacts which the various roads would have on the environment.

Approximately eighteen thousand copies of this brochure were delivered to households in the Newbury area by the Royal Mail. The brochure contained an invitation to attend exhibitions where drawings and photographs would be on display and where DoT and consultant's staff would be on hand to discuss informally the traffic, environmental or other issues of concern. Exhibitions were held in 1982 at the Council Chamber in Newbury District Council Offices on July 7, 8, 9, 10, 16 and 17 and at the Burghclere Village Hall on July 13 and 14.

A postage-paid questionnaire was included with the brochure seeking the public's views on the factors they considered to be of most significance to them in choosing a route. About 4,500 questionnaires and letters were received by the DoT together with two petitions. 57% of the respondents preferred the western route and 35% preferred a central route. About 25% asked that an eastern route should be given further consideration.

The newly-formed Society for the Prevention of the Western Bypass (SPEWBY), consisting mainly of residents living to the west of Newbury, submitted a comprehensive report supporting alternative eastern routes. Altogether, some 80 alternative routes or route variations were put forward. The four local Councils involved - Hampshire County Council, Basingstoke & Deane Borough Council, Berkshire County Council and Newbury District Council - all preferred the western route, but Berkshire and Newbury asked that an eastern route should be further investigated, including one across the Racecourse. The Nature Conservancy Council (now English Nature) proposed that a more southerly diversion at Snelsmore Common should be followed so as to lessen the effect of the bypass on the Common, which is a designated SSSI.

Publication and Description of the Western Route

It was not until June 1984 that the Minister of State for Transport announced the DoT's preference for a western route, substantially following the brown route described in the consultation brochure but diverted at the northern end both to reduce the effect on the Mary Hare School for the Deaf and to take into account the concern of the Nature Conservancy Council, so that only the southern corner of Snelsmore Common would be crossed by the route, taking 3 acres of the Common's then existing 250 acres. In its announcement, the DoT noted that it had not been possible to find a satisfactory eastern route and stated, *"The western alternative would affect agricultural land and sensitive environmental areas, which was regretted, but had least effect on people, the greatest public support and was best value for money."* The announcement also stated that further detailed design work would follow, including ecological and noise surveys, and that the provision of measures such as landscaping and screening would be considered. The publication of the western route enabled the DoT to protect the line by an Order under Town and Country Planning legislation.

The western route (see the Published Route line on the centre pages map of the area) is approximately 8½ miles, or 13½ kilometres, long. It commences at the southern end near Tot Hill by continuing the dual-carriageway from the overbridge which carries the Burghclere/Penwood road. The road then drops into a cutting with roundabouts provided on each side from which slip roads give access in both southerly and northerly directions to each carriageway. The road then curves towards the north-west to join the line of the dismantled railway, which it then follows for about three miles, skirting the woodland at Ball's Copse and various other plantations on an embankment before crossing over a minor road, The Drove and following the edge of Great Penwood.

Its intersection with the A343 Newbury/Andover road involves a re-alignment and depression of the A343 and the construction of slip roads in a field to the north of The Chase, which is an area of woodland owned by the National Trust. The road then skirts The Chase itself crossing over the minor road of Enborne Row and under Enborne Street, continuing to follow the line of the old railway through Reddings Copse in a cutting and over a re-aligned Wheatlands Lane on an embankment.

After passing Skinners Green Lane, the road leaves the line of the old railway to pass over Enborne Road and enter the Kennet Valley on an embankment with bridges over the London-Penzance railway, the Kennet & Avon Canal, the River Kennet and three other watercourses. The road passes under the A4 close to its junction with the B4000 Stockcross Road. There is a deep cutting at this point, with roundabouts on either side providing northerly and southerly slip road access to both carriageways. The roundabout on the west of the A34 also provides a junction to the B4000.

An embankment with bridges then carries the bypass over the Newbury/Lambourn road, requiring a re-alignment of the access road to Bagnor village, and over the River Lambourn close to Bagnor village itself. It is carried in a deep cutting within the Japanese-owned Donnington Grove golf course and bridges are provided to carry Footpath No 8 and Bridleway No 7 over the road. It enters a woodland area on an embankment and Snelsmore Common Country Park in a cutting, passing under the B4494 Wantage Road. It skirts the grounds of Arlington Manor, which houses the Mary Hare School, on an embankment and drops down to join the old line of the A34 south of the M4 in a long junction with slip roads requiring an underpass and a flyover.

Environmentally, it was clear that the most sensitive areas were those of The Chase, the Kennet Valley, Bagnor Village and Snelsmore Common. The National Trust were to become the principal objectors at the secondary inquiry in 1992 since they owned not only the Chase, but also the adjacent field in which the slip roads were to be built.

SPEWBY's alternative eastern route received wide coverage in the media. Alternative central routes also attracted attention and public opinion became polarised. In 1985, alarmed by the media exposure given to SPEWBY, the Newbury Bypass Supporters' Association was formed to lobby in support of the western route. The Queens Road Residents' Association in Newbury opposed any central route. Feelings were running high. Letters in the local paper became more emotional. Spewby posters in and round the town were countered by Supporters' Association posters which Spewby supporters were accused of vandalising.

The draft Orders initiating the formal procedures for the construction of the line of the western bypass were not published until 30 October 1986, nearly 2½ years after the DoT had announced its preference for a western route. Exhibitions, which included a topographical model, plans and display boards, were held in the Council Chamber of Newbury District Council on 6, 7, and 8 November 1986; at Burghclere Village Hall on 11 and 12 November and at Stockcross Village Hall on 14 and 15 November. In addition to the alternative eastern and central routes, the DoT received proposals for variations to the western route. With so many diverse views being expressed, it had become obvious that a major public inquiry was inevitable.

Evidence was assembled and lawyers briefed. The DoT was constantly pressed for an inquiry date. 1987 came and went. Eventually, in April 1988, starting dates were announced. The pre-inquiry meeting was to take place in May 1988 with the inquiry itself opening at Stockcross in June.

Newbury by-pass inquiry

<u>Groups get ready to present their arguments</u>

Battlelines drawn up for June inquiry

NEWS of the public inquiry date has been greeted with delight by Newbury's MP, with local groups anxious to present their views on the by-pass proposals.

Newbury MP, Mr Michael McNair-Wilson, said "I am delighted and relieved that at last we have got a firm date.

ity. We will be arguing for the western route."

Thatcham town clerk, Mr Peter Rusted, said "The council is opposed to the eastern route in principle. We have considered the options on several occasions. There was public consultation of the alternative routes and the town council opted

one because it goes through the countryside.

"We have chosen the eastern route as the best, and have worked on this. The advantage with an eastern route is it would link up with Thatcham traffic to ease the congestion.

"If it was possible to construct a good central route,

against the central route, open minded on the east, but prefer the western route.

"The eastern route is environmentally damaging, but the western route is also fairly damaging. The central route — is not on, because of the environmental damage, and there would be

County councillor for Cold Ash and West Thatcham, Commander Michael Porter, said "It must go west because it is far too late to go east or central. So much work has already been done on the western that it would be retrograde to try any other."

Newbury Chamber of

Headline in The Newbury Weekly News on 14 April 1988

12

CHAPTER 3

Purpose of the 1988 Public Inquiry

Some misconceptions exist about the purpose of a public inquiry into a road scheme. In the words of the DoT's booklet, "Public Inquiries into Road Proposals", the purpose of the public inquiry is, *"to inform the Secretaries of State for Transport and the Environment of the weight and nature of objections to a road scheme. The key tasks of the Inspector are to take account of objections from people affected by the proposals; to report on those objections; and to make recommendations to the Secretaries of State for Transport and the Environment on the proposals. The ultimate decision is not the Inspector's; it is one which the two Secretaries of State take jointly, in the light of representations and objections, the Inspector's report, and all relevant aspects of the Government's policies."*

The Inspector has considerable discretion as to the programme and times of the sittings and the dates of site visits. He considers the scheme published by the Secretaries of State. It is not his duty to choose the best scheme among those put forward at the Inquiry. He may recommend wholesale rejection of the proposed scheme on the grounds that an alternative seems preferable, but if his view is accepted by the Secretaries of State, then new procedures would have to be initiated. The DoT would begin the investigation afresh and would need to instruct its consultants to work up the new scheme, publish it and proceed to another inevitable inquiry. Such a course would clearly delay the project for several years.

A panel of suitably experienced and qualified Inspectors is held by the Lord Chancellor's office from which a nomination is made by the Planning Inspectorate. The choice of Inspector will depend upon the particular issues in the case. For the main public inquiry at Newbury, Lt Col M F Davies was nominated. Due to the complexity of the issues which were to be considered, he was attended and assisted by a technical assessor, Mr A Langton. The facilities and staff for the inquiry were provided by the DoT.

The pre-inquiry meeting was held on 11 May 1988 at Newbury Racecourse to enable the Inspector to hear from those intending to give evidence and to draw up a rough programme for the proceedings. The inquiry itself was opened on 28 June 1988 at Stockcross Village Hall where Portacabins had been sited to house the Inspector, the DoT staff, the consultants and lawyers and to provide library and conference facilities for witnesses. The large three-dimensional model of the Newbury area was on permanent display in the entrance lobby and large scale plans of the western route were displayed within the main hall.

Support for the Western Route

The first section of the inquiry was taken up by a presentation of the case for the proposed western route, referred to throughout as the "PR" or published route. This was led by the DoT's counsel, Mr Charles Calvert. The need for improvement was explained in detail and covered the extent of congestion and delay, accident rates, and the effect of the present road on the environment and on people and property, such as community severance. The impact of the proposed route on the ecology of the area was given, including that on the SSSI at Snelsmore Common and on wildlife habitats during construction.

Landscaping was also explained in some detail. Large artists' photomontages had been prepared showing the panoramas from two dozen key positions along the route. These gave 1) the views which then existed, 2) as they would appear on the day of opening of the bypass and 3) as they would appear when the extensive tree and shrub screening had reached maturity.

The results of traffic surveys taken since 1979 were given. Predicted traffic flows before and after the opening of the bypass had been studied in depth and the figures produced by computer modelling were produced and explained. The predicted noise levels were given and showed that with the western bypass, 187 dwellings overall would experience a noise increase and 573 a decrease.

General support for the DoT's scheme was expressed by witnesses from those Councils directly affected by it, namely Berkshire County Council, Hampshire County Council, Newbury District

Council and Basingstoke & Deane Borough Council, with some minor comments on the design features. (Berkshire CC also appeared as an objector against the Kennet Valley crossing. This was considered later in the inquiry - see pages 16/17).

Another supporter was the locally prestigious Newbury Society, whose membership was reported to the Inspector as having, *"the aim of helping to conserve the best of the old and maintain and enhance the Newbury environment in the future"*. The Newbury Bypass Supporters' Association also gave details of its objectives, constitution and membership and presented a petition of over 6000 signatures collected in the Newbury area in 1987 and 1988 expressing support for the DoT's route. The result of the petition also showed that 89% of those approached were in favour of the western bypass. The Association pointed out that the need for bypass was undisputed and that the continuing delay was causing considerable anxiety. Various individuals also expressed their support for the western bypass including local Councillors and the Rector of Burghclere and Newtown.

Objections to the Western Route

The National Trust objected on the basis that its land at The Chase, adjacent to the A343 Andover Road, would become noisier for visitors and wildlife as the bypass would pass along its northern boundary. If a western route were to be built, the Trust proposed that at this section the bypass should be at ground level, keeping the old railway embankment as a noise barrier ,and pass under the A 343, but the DoT pointed out that the Chase already suffered noise from the A343 and that it was doubtful whether the increase would have any significant adverse effects. On the other hand, the Trust's proposal would cause noise problems to the adjacent houses. In his report, the Inspector did not consider that the Trust's arguments carried sufficient weight to justify the change. The Trust were to return with lengthier objections at the second inquiry in 1992.

The actual need for a junction with the A343 Andover Road was questioned, but the Inspector reported that the forecast traffic justified it and that, *"it would appear to me to be wasting the potential of the PR to remove traffic from Newbury if the junction was omitted"*.

Objections to the western route also came from farmers in the Enborne area who understandably were concerned at losing land and accesses to their fields. One of these cases illustrates the independent and questioning attitude adopted by the Inspector. Boames Farm had tenants of long standing who ran a strong Guernsey dairy herd. When the railway had been constructed in 1885, a bridge had been provided over which the cows could be taken for milking. With the replacement of the railway by the wider bypass, the DoT were not proposing to replace the bridge. The Inspector wanted to know why, and the DoT representatives replied that it would cost £212,000 (at 1987 prices) for a bridge designed to take heavy farm equipment. A detour via a nearby minor road was feasible as an alternative. The farmer replied that he wanted a bridleway bridge for cattle alone. On the minor road, cattle herded four times each day past houses, whose gates might be left open, would require so much additional attendant manpower as to make his farm uneconomic. The Inspector asked for a new estimate for a bridleway bridge which would satisfy the farmer; this was eventually produced as £120,000. In his report, the Inspector commented that he found it *"totally unreasonable"* to expect the farmer to put up with the detour, that the expenditure *"can be justified on personal humanitarian grounds"* and that, *"I have no hesitation whatever in recommending that a bridleway crossing should be provided"*. The Secretaries of State agreed with his recommendation and the bridge was approved.

Snelsmore House objected to the visual and noise intrusion which would detract from its use as a centre and retreat for the Order of the Cross, an informal Christian Fellowship. However the Inspector regretted that he could not recommend either the change of line or the cut-and-cover tunnel requested by the Order and, as with some other objections, the Inspector could only say that, *"any residual detrimental effects must remain a matter for compensation"*.

The parish of Speen includes the villages of Speen, Woodspeen. Marsh Benham, Stockcross and Bagnor, the latter being most affected by the proximity of the bypass, and Speen Parish Council objected to the route which they considered would cause additional hazards and accident black spots.. However, the Inspector said that he must be guided by the official figures provided for accidents and traffic which indicated no serious problem. Neither could he agree that the bypass would cause severance within the parish which would adversely affect the way of life of local people nor a threat to the viability of farms in the parish.

Miss Helen Anscomb, a local campaigner against road building, questioned the need for the bypass and also raised points which included the injustice of road building, the unsatisfactory public inquiry procedures, women drivers and driving licences, drinking and driving, speeding, and the reduction of traffic which would occur on the opening of the Channel Tunnel. The Inspector said that he understood her frustration at finding no forum where her general views could be considered but mostly they dealt with matters outside the scope of the inquiry. However he made a record of her evidence and drew the attention of the Secretaries of State to it. They said, in their later decision letter of 1990, approving the Inspector's recommendations, that the matters were not appropriate to a public inquiry on a specific road proposal but that, *"it is open to anyone to make representations about the Government's transport policy either direct to the Department or through their elected representative at any time."*

In subsequent evidence at the inquiry, the Bypass Supporters' Association said that the traffic to the docks at Southampton, Portsmouth and Poole constituted the largest source of HGVs on the A34 at Newbury. They had contacted the docks managers of those ports who had said that their European commercial traffic generally came from the Midlands and the North and had destinations to southern and western France or the Iberian peninsular and would be unlikely to use the Channel Tunnel. In the case of the principal port, Southampton, continued expansion of the port facilities would lead to an increase in docks traffic of which the majority was "long sea", i.e. heading for destinations beyond Europe. Southampton is the largest UK container port and the major outlet for car exports. The subsequent opening of the Channel Tunnel appeared to confirm the forecast that it would have negligible effect on the A34 traffic at Newbury.

Of the 147 witnesses appearing at the inquiry, only two queried the need for improvement of the A34. The Secretaries of State in their 1990 decision letter said, *"The Inspector concluded that Newbury did need a bypass and the sooner it could be built the better it would be for the people of Newbury and the users of the A34. The Secretaries agree with the Inspector's conclusion."*

CHAPTER 4

Alternative routes (See area map on centre pages)

Objectors had known of the DoT's western route preference since June 1984 and therefore had four years to prepare their alternatives. On April 28 1988, the DoT had placed an advertisement in the Newbury Weekly News, the lively and dominant local paper, founded in 1867, which has a circulation of over 27,000 throughout west Berkshire and north Hampshire. The advertisement showed plans of the objectors' alternatives, drawn up by the DoT's consultants to a uniform scale of approximately 1/25,000. Eight of these were variations to the proposed western route, ranging from substantial re-alignments of the southern half to the provision of extra bridges and a viaduct. Two were routes to the east of the racecourse, through Thatcham, and one was a route east of the town centre but crossing the western end of the Racecourse. Six others were alternative routes through the centre of the town.

The Variations to the Western Route

The most significant variation, WA3, was a proposal to move about three miles of the route away from the old railway and north-eastwards into the countryside at Wash Water and the Enborne Valley. This would reduce the bypass length by half a kilometre and give a cost saving, according to the DoT, of £1.05million at 1987 prices. Nevertheless the DoT were opposed to it because of the adverse effects it would have on the landscape and to agriculture and the Local Authorities were also of that opinion. Local farmers opposed the change because of the severance it would inflict upon their properties. The Inspector's view was that the economic benefits were, *"insufficient to make up for its visual shortcomings"* and concluded that it should not be adopted.

The decision on another variation at the southern end of the bypass proved to be more finely balanced. Residents at Tot Hill put forward a line about 1½ miles long, WA5, which would move the bypass south-westwards, further away from the old railway and away from their houses but into a woodland designated an Area of Outstanding Natural Beauty (AONB). This would require a shallower cutting than the PR and would give a cost saving of nearly £1million. The Inspector recommended that this variation should be accepted but the Secretaries of State considered that the intrusion into the AONB was not justified and that such a major alteration to the line would require new statutory procedures which would cause serious delay to the scheme. It was accordingly rejected.

WA6, WA7, WA8 and WA13 related to junction modifications which were rejected. Perhaps the most significant of these was WA8, which would have provided an extra and convenient junction for Newbury town on the Enborne Road but was turned down because it would have put more traffic on to unsuitable residential roads.

WA10 was a proposal by Berkshire CC and Newbury DC, supported by the Ramblers Association, for a footbridge at Skinners Green instead of the "at grade" crossing which the DoT were suggesting following surveys which indicated that very few walkers used the adjacent footpaths (Nos 5 and 6A). In spite of the additional £70,000 cost of the footbridge, the Inspector's strong recommendation to provide it was accepted by the Secretaries of State who noted that the route was being published in a leaflet and that an increased use of it might result. (Following a request at the 1992 inquiry, this footbridge was moved and upgraded to an accommodation bridge for agricultural use.)

The diversion of the bridleway linking Donnington Castle and Snelsmore Common was challenged by the Speen Parish Council. The DoT said that to maintain the existing line would require a long skew bridge across the bypass costing £240,000 more than the shorter bridge being provided. Since the diversion lengthens the bridleway by only 200 metres, the Inspector said he could not recommend the Speen proposal.

The Kennet Valley Crossing

A major change to the Kennet Valley crossing, WA9, was proposed by Berkshire County Council who for some years had wanted to see a long viaduct for this section instead of the proposed embankment with its five bridges. The Nature Conservancy Council (now English Nature) also favoured a viaduct. Flooding of the meadows upstream of the proposed crossing is a constant occurrence but the DoT's consultants did not consider that the infrequent increase in the flooded area caused by an embankment would cause special difficulties and would certainly not affect the railway embankment. Expert evidence was produced on the ecology of the valley and various aspects were debated in detail. Translations of recent German work were produced indicating that one significant species, the forest ground beetle, would be unlikely to crawl across the surface of the dual-carriageway bypass, but the experts present were unable to answer the Inspector's question as to why, being a beetle, it could not fly across. The embankment would be provided with shallow slopes allowing either cultivation or tree planting to be carried out up to the carriageway fence itself. The DoT's expert argued that a viaduct would create a *"channel of darkness 180 metres long and 30 metres wide"* which would inhibit the movement of various animals whereas, with an embankment, *"birds, butterflies and dragonflies should cross the carriageway in the light."*

In his conclusions on the road v. viaduct landscaping issue, the Inspector acknowledged that both solutions would be intrusive and, *"it is very much an individual perception as to which is preferable"* but he did not favour the viaduct *"as it could never mellow nor merge into the landscape in quite the same way as an embankment."* The Inspector found that the ecological evidence was evenly balanced and inconclusive and could not be considered to give weight to either alternative. However he was influenced by a request from the Ramblers Association who, in their evidence, had asked for a footpath to be provided alongside the bypass to link the Newbury/Speen Footpath No 2 to the canal towpath, thereby providing a pleasant circular walk from Newbury town centre. This could be incorporated along an embankment more readily than on the edge of a viaduct and the Inspector *"strongly recommended it"*. In the light of the Inspector's conclusions, it is not surprising that the Secretaries of State considered that the additional £5.8million cost of a viaduct could not be justified.

The National Rivers Authority, formed since the inquiry, expressed concern about the restriction which the embankment would cause to flood water. During the final design work on the bypass, with the Rivers Kennet and Lambourn being designated as SSSIs, the Highways Agency announced on 22 February 1996 that two bridge spans in each of those valleys would be widened.

The Proposed Eastern Routes
SPEWBY Route EA1A

The principal objection to the western route came from SPEWBY, a majority of whose members lived in the area which would be affected by it. At the inquiry, they were represented by an experienced counsel, Mr J Burford Q.C., and junior counsel, Mr R Langham, and by expert witnesses on traffic, landscape and environmental matters. It seemed that SPEWBY possessed financial resources considerably greater than the normal subscriptions collected from its 330 members and the Bypass Supporters Association asked for these sources to be disclosed in view of the large proportion of the inquiry's time which would be taken up in the presentation of their case, but the Inspector said that he did not need to know of the origins of their funding. The SPEWBY chairman commented that, *"I don't think I can talk about our financing . . . there is a lot of confidentiality involved."*

SPEWBY objected to the effect that the western route would have on the environment and the landscape and made much of a report by the Department's own Landscape Advisory Committee which had condemned any western route (but which had not been asked to consider the impact of the alternatives). However the Inspector said, *"I view the conclusions of the LAC with great respect, but I have to consider them in the context of all the other evidence that I have heard during the Inquiries."* SPEWBY had proposed the eastern alternative designated EA1 but, during the inquiry, toxic waste was discovered in a tip over which EA1 would pass. Accordingly the Inspector extended the summer recess in the inquiry to allow SPEWBY to modify their route, afterwards designated EA1A. The recess lasted from 15 July 1988 to 6 September 1988.

From its southern end, route EA1A would generally follow the existing A34 along the 1½mile section known as the Newtown Straight, through Sandleford Park and having a junction at Pinchington Lane. The route would then swing eastwards between Greenham Common and the Racecourse with a junction on the A4 London Road. It would then turn westwards passing north of Shaw to rejoin the A34 on the Donnington Link north of High Wood.

The SPEWBY route would be more convenient for goods vehicles having their origin or destination in Thatcham and for traffic using the A339 from Basingstoke, although these together form a small proportion of the total using the A34. EA1A would be longer than the existing A34 through Newbury by about 2.8 kilometres or 1¾ miles. In addition, the long tight curve round the Racecourse would require a permanent 50mph speed limit. It was therefore accepted both by SPEWBY and the DoT that some traffic, particularly at off-peak times, would continue to use the existing A34 rather than the longer eastern bypass, although there was a difference of opinion as to the percentage which would do so. The Inspector concluded that doubt existed on the likely effectiveness of the SPEWBY route.

The Newtown Straight section at the southern end of the SPEWBY route proved particularly contentious. There are a surprising number of residents along this section; the Bypass Supporters' Association reported that 257 were listed on the electoral roll compared with 55 at Bagnor, the residential area most affected by the PR. The dual-carriageway would therefore need a central reservation sufficiently wide to accommodate pedestrians waiting to cross, or be provided with footbridges. Service roads would be required on each side of the main road to give access to the numerous lanes and driveways. The Supporters' Association also pointed out that 2-metre high anti-dazzle fences would be necessary between the bypass and the service roads as well as crash barriers in the central reservation and that the total width of construction would accordingly be over 46 metres, compared with the existing road plus footpath construction of about 8 metres, thereby taking a considerable strip off Newtown Common. The Association was allowed to cross-examine SPEWBY's landscape consultant on his view that such an arrangement, *"would provide a very fine approach to Newbury from the south"*. The Inspector reported, *"It is quite clear to me that those who live there do not subscribe to this view, neither does the Department and I certainly do not"*.

A major effect of this widening of the Newtown Straight would be the effect on the 30-acre Quarry Wood garden which is *"internationally recognised"* for its area of exotic woodland which provides a microclimate for rhododendrons, azaleas, magnolias, acers and, in particular, for a range of lilies, many of which have become rare in their native habitats. The garden is privately owned by Mr & Mrs Martyn Simmons, the latter being a leading expert on lilies. The SPEWBY proposal would have required the felling of many of the mature trees which give the garden its unique character. The National Council for the Conservation of Plants and Gardens appeared at the inquiry, supported by letters from eminent horticulturalists, including the Director General of the Royal Horticultural Society, to oppose this section of the SPEWBY route. The Inspector paid a formal visit to the garden accompanied by DoT, SPEWBY and Supporters' Association representatives. He subsequently reported that, *"the effect on Quarry Wood in particular, and Newtown Common in general, must go in the balance against EA1A"*.

After leaving the Pinchington Lane junction, the SPEWBY route would pass close to a residential area. According to the DoT, 244 houses would be within 100 metres of EA1A compared with 48 for the PR. SPEWBY had criticised the PR for its conspicuous embankment across the Kennet Valley but the SPEWBY route would also require a long embankment and the Inspector's opinion was that, *"in general terms both roads would be equally intrusive within the two valleys and neither any more or less acceptable than the other"*. The SPEWBY crossing was attacked by the DoT, by the local councils and by the Supporters' Association for the damaging effect it would have on the Thatcham Reed Beds which is, *"one of the largest inland reed bed complexes in southern England and is of particular interest for its resident and migrating population of birds and for its entomology"*. It is used by large number of schools and other organisations. Evidence was given that SPEWBY's viaduct in this area would damage its sensitive hydrology. In view of these facts, and the building of an excellent visitor and educational centre by Newbury District Council with help from the Royal Society for the Protection of Birds, it is surprising that the RSPB itself was not represented at the inquiry in 1988 to help in defending this important location.

The Berkshire, Buckinghamshire and Oxfordshire Naturalists' Trust (BBONT) considered that all of the bypass routes were damaging to nature conservation but the eastern routes were the worst in this respect due to their effect upon Thatcham Moors. The Nature Conservancy Council concluded at the end of their 83-page evidence that, *"The construction of route EA1A through Thatcham Reedbeds would have a serious deleterious effect on the habitats of the Local Nature Reserve, and on the range of wildlife species present, their populations and their enjoyment by the public. Accordingly, NCC considers that this scale of damage is unjustifiable and submits to the Inspector that the objectors' eastern route should be rejected."*

The SPEWBY route would require a major junction at its intersection with the A4. The DoT reported that, *"Here the elevated stretch of road would be some 1.5km long and this, together with the two roundabouts and local access roads, would be the most intrusive feature of the whole route."* The future traffic problems at Lower Way, Thatcham, with the build up of local traffic was seen by the Inspector as, *"a major failure in EA1A . . .simply because the predominant traffic flow is east-west."*

The cost of SPEWBY'S route was put at £80.85M as opposed to £46.65M for the PR, or some 73% greater.

Routes EA2B and EA3A

Both of these routes followed a similar line to the SPEWBY route along the Newtown Straight and were condemned by the Inspector for the same reasons. EA2B followed the tight bends of the SPEWBY route round the Racecourse and, instead of a viaduct, crossed the Thatcham Moors on an embankment which the Inspector said, *"probably does more damage to the area"*. EA2B then kept farther to the east than the SPEWBY line, requiring the demolition of more houses - 76 in total, 56 being in the Lower Way area. Its greater length than the SPEWBY route led the Inspector to comment, *"my doubts as to its effectiveness must be even stronger than they are for EA1A"*.

Instead of passing through Thatcham, EA3A would pass beneath the western end of the Racecourse in a cut-and-cover tunnel with another tunnel under Faraday Road and a junction with the A4 at the Robin Hood roundabout. The complex engineering structures required were criticised by the DoT and the cost, at £111.73M (140% greater than the PR) caused the Inspector to comment, *"I simply do not consider that this additional cost can be justified in any way whatsoever"*.

The Proposed Central Routes

The route designated CA1 was put forward by the well-respected consultants, Travers Morgan, on behalf of the Council for the Protection of Rural England (CPRE). It took a line to the west of Tot Hill and Horris Hill and passed through Sandleford Park. A new junction was provided at Pinchington Lane. The dual-carriageway section through the centre of the town had a southern access at St John's Road and a northern access at the Robin Hood Roundabout with northbound and southbound roads for local traffic on the west side and east side respectively, requiring considerably more land than the existing road and appreciable demolition of property. Both carriageways of the bypass would be carried on flyovers at St John's Road, Kings Road (Sainsbury's roundabout) and the Robin Hood Roundabout.

The CPRE route had been expertly prepared and had the merit of avoiding the destructive dual-carriageway and service roads through the Newtown Common area which other alternatives were proposing. It could only do this by providing a new route through an environmentally sensitive country area within Hampshire across the Enborne Valley and Falkland Farm. But its major drawback was the destruction of domestic property that would be necessary within Newbury town centre and the difficulty of constructing the road while allowing reasonable passage of A34 traffic. These severe disadvantages became apparent to the CPRE's representatives during the detailed cross examination of their consultant's engineer and they requested a short adjournment, after which they asked for their proposal to be withdrawn. The Inspector acknowledged that their decision must have been a difficult one in view of the work put into its preparation. However, he ruled that, in the light of the support expressed for the central routes, the DoT's complete response to the proposal should be heard and said that he would fully report on it.

19

CA2 was routed along a dual two-lane carriageway on the Newtown Straight with continuous service roads on either side, similar to the SPEWBY proposal and with equally damaging consequences. It passed through Sandleford Park with a junction near Pinchington Lane. Queens Road was lowered through an underpass and St John's Road was linked to Cheap Street, crossing the railway. Kings Road was partially lowered through an underpass and at the Robin Hood Roundabout the bypass was carried on a flyover with exit and entry slip roads providing a junction with the A4.

CA3 had a dual carriageway along the Newtown Straight. Its sponsor hoped that the upgrading of existing tracks would give access to the numerous houses in the area. It passed through Sandleford Park with a junction at Pinchington Lane and underpasses at Queens Road and Kings Road and a flyover at the Robin Hood Roundabout. Junctions would be provided at Kings Road and the Robin Hood Roundabout.

Both CA2 and CA3 suffered from similar problems to CA1. They would require considerable demolition of property within Newbury. The flyover structures would be visually intrusive and noise and pollution problems would remain. As with CA1, the construction problems in the town would be immense and, when finished, a permanent 50 mph speed restriction would exist on the A34 due to the proximity of property.

CA4 had a dual carriageway along the Newtown Straight with similar continuous service roads as the Spewby route. It provided access to the A339 to Basingstoke via a modified Swan Roundabout. The roundabouts at Queens Road and Kings Road were unchanged, although the latter had traffic lights (since installed). A flyover was provided at the Robin Hood Roundabout with north and southbound slip roads.

CA5 was withdrawn. CA6 was not used.

CA7 was presented by Mr J Griffin, FRIBA. It had a similar design for the Newtown Straight section as CA4. It passed through Sandleford Park with a junction at Pinchington Lane and had a long elevated section through the town centre from south of Queens Road Roundabout to north of the Robin Hood Roundabout with southern access at Queens Road Roundabout and northern at the Robin Hood. Below the elevated section, the Inner Relief Road would remain for local traffic. (The design would therefore resemble the Chiswick flyover in west London.)

CA8 treated the Newtown Straight in a similar way to CA4 and CA7 and also provided a junction at Pinchington Lane. The bypass was then put in a one-mile tunnel from south of the Queens Road Roundabout to north of the Robin Hood Roundabout with southern access at its southern end and northern access at its northern end.

There would be extreme difficulty in constructing these routes both within the town and also along the Newtown Straight while trying to maintain the normal traffic flow along the A34. The cost of any effective central route was approximately twice that of the western bypass, the tunnel being more than four times as much.

(There are costs other than construction costs to be taken into account. The land purchase costs quoted at the inquiry were those of the District Valuer's in 1987 and, no doubt, they have altered appreciably. The savings due to reductions in journey times, accidents and vehicle operating costs over the 30-year assessment period were calculated according to the DoT's cost benefit programme in use at the time, COBA 9, and indicated greater benefits for the PR. Changes in these costs since 1988 are unlikely to have any influence on the relative benefits of the PR over the other proposed schemes.)

In the introduction to his Conclusions, the Inspector had said, *"When considering the various routes which have been proposed for a Newbury Bypass, I believe that there are just three prerequisites which must pass the test before a route should be seriously considered and all other factors then brought into consideration. These are, firstly, whether or not the route is likely to be effective in terms of traffic - in other words, does it achieve the aims? Secondly, the cost - is it viable in terms of NPV (net present value) and are the costs in terms of £M reasonable? Thirdly, is the impact on the environment, both human and rural, acceptable?*

Referring to the Central routes, the Inspector said, *"I have decided to consider the Central Routes together because of their similarity in what I have termed the three major prerequisites."* In terms of traffic effectiveness he said, *"I have not a shadow of a doubt that they all fail by a considerable degree to pass the traffic prerequisite."* He also said, *"In terms of cost, all the Central Route proposals fail to match the PR within anything like reasonable bounds except for CA4, but that is virtually ruled out by the Department as having severe and insurmountable operational difficulties."* On the environment issue, *"The Central Routes most certainly fail the human environment prerequisite."*

His additional opinion on the Central Routes was probably the most damning; *"Perhaps the most significant feature of these Central Routes would be the heinous upheaval and disruption that would take place during construction, some routes worse than others, and then when all of that was over, Newbury would, in the case of most of the Central Routes, be left with constructions totally out of keeping with a country market town."*

The Friends of the Earth did not attend the 1988 inquiry. If they had done so, they might have saved themselves the expense of promoting their own On-Line Alternative central route in 1994, for which the Inspector's comments would doubtless have applied with equal relevance.

General Comments on the 1988 Inquiry

The 1988 inquiry was continued by the Inspector until all of those wishing to give evidence had done so. Although the inquiry was normally open during the working day from Tuesday to Friday (Monday being reserved for site visits), the Inspector had stated on several occasions that he was willing to hold sessions at other times to accommodate witnesses, although only one evening hearing was actually necessary. There had been no question of anyone being excluded from the investigation process. One hundred and forty-seven witnesses had given evidence.

Newbury's Conservative MP at the time, the late Sir Michael McNair-Wilson, gave evidence saying that he was generally in favour of the western route but that the evidence was that 10-15 years after completion of the bypass the town would be just as congested unless local road improvements were carried out. In his conclusions the Inspector stated, *"A frequent criticism of the PR is that it would not solve all Newbury's traffic problems. However the Department has never pretended that it would. What it would do, unlike any of the alternatives presented, would be to provide sufficient relief up to 2009 (the design year) to enable local authorities to introduce measures to relieve the residual problems."* The myth that the DoT were claiming to solve both local and through traffic problems by building the bypass persisted for some years. Though totally without foundation, it was frequently reiterated by the road protesters.

About one third of the bypass at its southern end lies within the north-west Hampshire constituency whose Conservative MP was then Sir David Mitchell. The Bypass Supporters' Association urged him to attend the inquiry but, although he expressed support for the Department's route, he declined to become involved saying that, because it was less than 12 months since he had resigned his ministerial post within the Department of Transport, protocol precluded his taking sides in a DoT project. However he was to play an active part later in getting the bypass restored.

Perhaps the most important admission to emerge from the Inspector's report was the change in his thinking during the course of the inquiry. Like many observers from outside the Newbury area he admitted to an initial prejudice against the western route; *"Prior to the Inquiries the Assessor and I walked most of the line of the PR and we were lucky enough to choose two days of the English high summer at its best; this left us with a deep impression of the beauty and variety of this stretch of countryside more lasting than any evidence given during the inquiries could possibly do. Our joint reaction was that there would have to be very good justification from the Department of Transport for driving a new dual carriageway road through such scenery."* His principal duty was expressed in his Summary of Conclusions, *"I am aware that the most contentious aspect of the PR is the effect it would have on the landscape; throughout the Inquiries this has been uppermost in my mind."* He went on to say, *"As the Inquiries progressed and I heard the evidence from all sides, I found that the Department's case became stronger and more convincing, and I am now entirely satisfied that the adverse effect that there would be on the western landscape is justified by the effect that any alternative would have on people and property to the east."* Some erstwhile protesters have already admitted a similar change of heart as they became aware of the facts.

One of the Inspector's duties was to gauge public reaction to the proposals. In addition to the Bypass Supporters' Association's 6000+ signature petition, two others were tabled at the inquiry, one of 700 signatures supporting the PR and another, of over 900 signatures from the Queens Road Residents Association opposing all central routes. Taking into account also the support received from the two County Councils, the two District Councils and various witnesses, the Inspector stated, *"I can only conclude that there is overwhelming public support for the PR".*

The inquiry held its final session on 9 November 1988. Seven months later, in June 1989, the Bypass Supporters' Association, reflecting local concerns, were pressing for a decision through Sir David Mitchell MP, but the Minister for Roads, Peter Bottomley, replied on 27 June 1989 that, *"In view of the complex nature of this major scheme, an announcement cannot be expected much before the turn of the year."* One year on, in June 1990, the Supporters' Association was describing the continuing delay as a scandal.

The Department's decision to accept the Inspector's recommendation for the western route was not announced in Parliament until 19 July 1990, more than 1½ years after the completion of the inquiry. The long delay has never been satisfactorily explained. The Inspector's recommendations were positive and unequivocal. He would have partly prepared his report in the 7½ week hiatus in the middle of the inquiry and it is likely that the Secretaries of State would have received it early in 1989. In the interim, frustration and speculation had been expressed in the letters columns of the Newbury Weekly News and belated proposals had began to emerge for central routes and tunnels from proposers who had clearly forgotten, or had not followed, the proceedings of the inquiry. These continued for some years after the decision was made and from time to time attracted sufficient support to gain limited public credence and prominence in the media. The DoT was not prepared to refute the claims made for these later alternatives but the Department's delay and silence seemed to be sowing doubt in the public's mind about the likelihood of the western bypass ever being built.

A formal letter was issued, dated 16 July 1990 setting out the decisions of the Secretaries of State on the various points recommended by the Inspector and copies of the Inspector's 488-page report were made available at key locations in and around Newbury. The cheapest route, incorporating WA3, had been rejected as causing too much environmental damage. One of the Inspector's recommendations, for WA5, giving another cheaper route, had also had been rejected on environmental grounds. The Line Orders were formally "made" in 1991. The setting-up of the inevitable secondary inquiry into the compulsory purchase orders and other matters was then awaited. And awaited. It did not take place until 3½ years after the end of the first inquiry.

CHAPTER 5

The 1992 Public Inquiry and The National Trust

Draft Orders had been published from 4 February 1988 onwards by the DoT to cover issues outside the main line of the bypass. These covered the stopping-up or diversion of various lengths of roads, footpaths and private accesses, the provision of slip roads, temporary diversions, land purchase, work on watercourses and a certificate to provide land to Newbury District Council at Snelsmore Common in exchange for the comparatively small area being taken by the DoT. The normal secondary public inquiry into these matters was convened on 24 March 1992 at Stockcross Village Hall, the Inspector being Sir Peter Buchanan.

Berkshire County Council attempted to use the inquiry to re-open the debate on the viaduct v. embankment crossing of the Kennet Valley but the Inspector agreed with the DoT that that particular issue had been fully dealt with at the 1988 inquiry and ruled it out of order.

There was some dismay expressed that the bypass would not provide a route for exceptionally high vehicles but the DoT said neither did the M4 and M40, and the Inspector eventually recommended that the bypass bridge clearances should be of standard height.

The main issue at the inquiry was an objection by the National Trust to the acquisition of an open pasture which it owned adjacent to "The Chase" which is a woodland nature reserve and arboretum to which the public have access. The pasture itself, over which there was no right of way, was required for the construction of slip roads for the A343 Andover Road junction. The Trust was legally represented and its witnesses were the Regional Director and its Land Agent, based at Polesden Lacey, near Dorking. It did not appear that local National Trust members had been asked to give evidence. A large part of the Trust's proof of evidence was taken up by the expression of a grievance which it had with the DoT over the late replies it had received to correspondence; a matter which the Trust could probably, through its contacts, have taken up earlier at Ministerial level rather than absorb time at a public inquiry, whose Inspector obviously had no locus in such a dispute.

The important aspect of the land forming The Chase and its adjacent pasture was that it had been partly gifted, in 1944, and partly bequeathed, in 1977, by Sir Kenneth Swan and both pieces of land had been declared "inalienable" under the National Trust Act of 1907. The significance of the inalienable label is that it enables any attempt to compulsorily purchase the protected property to be resisted by conferring on the Trust the right to appeal directly to Parliament.

As an alternative to the design of junction proposed by the DoT, the National Trust put forward a proposal, OA2, (at WA2) for a layout of slip roads to the south-east of the A343 within an area of woodland which it did not own. The Trust's design would also bring the main A343 closer to a nearby row of houses than the DoT's design and involve a higher embankment for the bypass. The Trust had apparently made no effort to discover the views of any local members to its alternative junction design.

During the inquiry, the Bypass Supporters' Association discovered that the Trust had also informed neither the owner of the woodland nor the owners of the affected houses of its 12-month-old, but unpublished, alternative. At short notice, the woodland's owners rushed to employ professional help to represent them at the inquiry to fight off the Trust's threat. The house owners were understandably indignant that the service road and tree-screening provided by the DoT, and on which they had been consulted, might be replaced once again by the main A343, together with a higher bypass.

Although the acid pasture had for 25 years been periodically fertilised and rented out for cattle grazing, the Trust's representatives stated that it could readily be turned into a SSSI, a claim which, to many attending the inquiry, stretched credibility too far. After hearing ecological evidence, the Inspector's conclusion was that, *"There is no doubt that it is not of SSSI or any other protected status and it is unlikely to become so"*. The Department, on the other hand, in its scheme were intending to remove the existing calcareous road embankment supporting the A343 along that section, thereby helping the area to retain its acidic nature. The Trust's design involving night-long illuminated roundabouts was described by the Inspector as, *"environmentally intrusive in this rural setting"* and would cost nearly £½M more than the Department's design.

The Trust's lawyer referred to the EEC Environmental Impact Directive (see Chapter 8), presumably hinting that the EEC might be approached if the Trust was unsuccessful, and concluded, on the instructions of the Trust's legal department, with a request for an adjournment to allow a complete re-assessment of the junction to be undertaken, while claiming that the Trust had no wish to delay the scheme. Ignoring protocol, the DoT's lawyer was allowed to question the Trust's lawyer as to how those two diverse intentions might be reconciled, but he received no clarification. The Inspector, expressing his view that they were incompatible, pointed out that, *"should the Secretaries of State be disposed to accept OA2 . . . . I consider a third public inquiry, following publication of new draft Orders to be almost inevitable."*

An adjournment of the inquiry was, however, made from 30 April to 16 June in order for discussions to take place between the DoT and Newbury District Council over the exchange land certificate in respect of Snelsmore Common. The three areas given in exchange were eventually agreed after conclusion of the inquiry.

The decision letter of the Secretaries of State was published on 24 March 1993 and generally supported the conclusions of the Inspector. The National Trust's junction design was rejected. A number of other minor changes were resolved or were left to be dealt with by compensation or by further discussions between the Department and local landowners.

The Trust persisted with its opposition after the inquiry although the Inspector's recommendations and the Secretaries of States' decision had firmly gone against them. There is no doubt that the Trust had become exasperated with the acquisition of its inalienable land elsewhere. It had not used its right of appeal to Parliament for more than 20 years. The Southern Region Land Agent was quoted in The Guardian newspaper on 30 March 1992 as saying, *"If we let the DoT keep chipping away at estates given in perpetuity, then the public will lose faith in us. We have asked for the backing of the national council of the Trust to make a stand and we have been given it".* Another spokesman was reported as saying, *"We are prepared to go all the way on this one"* and that, *"We believe we now have a much bigger spanner to throw in the works than before and we intend to throw it."* Some parts of the quotations were denied by the Trust but emphatically confirmed by the reputable reporter concerned. Whatever the actual wording, the Trust's attitude was clear enough. It was going to take a robust stance over the Newbury Bypass.

There was virtually no chance that an appeal to Parliament would be successful in the light of the Inspector's conclusions, but the lengthy process of the Parliamentary procedure would be almost certain to delay the bypass by at least one year. The DoT could not continue with its procedures to build the bypass while there existed the possibility of Parliamentary intervention, however remote.

The threat was being used as a lever by the Trust to obtain the measures it wanted for noise protection and screening for The Chase itself. A letter from its Southern Region office to a Newbury resident in July said that the DoT had been asked by the Trust for mitigating works and stated, *" If they appear to go as far as they can, the Executive Committee have decided that it is minded not to object further to the compulsory acquisition of this inalienable land".* The Supporters' Association found it incomprehensible that such a prestigious organisation should act in what was seen to be an antisocial manner towards the people of the Newbury area and regarded the Trust's action as a misuse of its powers, but the Trust's Regional Director himself refused to meet or discuss its actions and their effects with the Association's representatives.

With considerable reluctance, (particularly since many of its members were doubtless also members of the National Trust) the Supporters' Association decided to compile and release to the Press, via the Newbury Weekly News, its detailed views on the way the Trust had conducted its campaign and set the Trust a deadline to retract its opposition. Eventually, on 4 August 1993, just before the deadline, the Regional Director telephoned, not the Association, but the Newbury Weekly News to say it would not be pursuing an application for parliamentary procedure. This action allowed the DoT to continue with the formalities for the construction of the bypass.

CHAPTER 6

Delays and Opposition

The article below the headline in The Newbury Weekly News on 5 August 1993 stated, *"The last major hurdle holding up the Newbury Bypass was cleared yesterday when the National Trust finally withdrew its objections to the scheme"*.

The Newbury Bypass had not been listed in the 93/94 Roads Programme, published in February 1993, but even so an early start was expected. The main contract work was unlikely to begin before the start of the next financial year in April, 1994, but the preliminary woodland clearance work was expected to commence during the winter season.

/bury Weekly News

THURSDAY, AUGUST 5 1993 Price 30p

IAM Theale Tadley **KINGSCLERE** Woolton Hill Burghclere **HUNGERFORD** Kinbury Inkpen LAI

ROUTE ALL CLEAR FOR BY-PASS

REPORT BY
STEPHEN ROUSE

THE LAST major hurdle holding up the

A start date during January 1994, predicted by its consultants, could not be confirmed by the DoT. Complaints were being made by residents outside Newbury that their villages were being used as "rat-runs" by drivers attempting to avoid the A34. In September, two crashes within four days on the Newtown Straight, described as *"horrific"* in the local press, pointed to the urgency of an early start and the DoT was constantly pressed for a date. The Compulsory Purchase Order for the bypass land was published on 7 October 1993 but on 4 November, the Newbury Weekly News headline read, "NEWBURY BY-PASS IN DANGER". The DoT was quoted as saying that the scheme would begin, *"when funds were available"*. It was thought that the fate of the bypass would have to await the publication of the Roads Programme for the year commencing 1 April 1994. David Rendel had been elected the Liberal Democrat Newbury MP in a by-election in May 1993 and he asked questions of Ministers in the House of Commons whenever the opportunity arose. In December 1993, there was confirmation that the Government was intending to delay the project.

NEWBURY Hermitage Cold Ash **THATCHAM** Theale Tadley **KINGSCLERE** Woo Burg

NEW BY-PASS DELAYED YET AGAIN

REPORT BY
STEPHEN ROUSE

The Newbury District Council had been carrying out a contract on behalf of the DoT to sound-proof some houses adjacent to the route. On 16 December, the Newbury Environmental Health officer was quoted as saying, *"We have stopped work on sound insulation because the DoT is waiting for confirmation of the timing of the scheme"*. A few weeks later, the Roads Minister told the NW Hants MP, Sir David Mitchell, that ground clearance work had been called off.

Then, on 28 February 1994, the Secretary of State for Transport, John MacGregor, announced in the House of Commons that funds for the bypass were to be made available after all. The Roads Minister, Robert Key, said the Government had responded to, *"enormous public demand"* for the road, adding, *"I hardly know anyone who has not been stuck in a traffic jam around Newbury."* However, the delay on funding was to be compounded since, under wildlife and countryside legislation, the woodland clearance work could not now start until the end of the bird-nesting season towards the end of August.

By then a new threat existed in the woods - tree-climbing protesters.

Iry Weekly News

NATIONWIDE CARPETS
SEAGRASS
SISAL
COIR
THE ALTERNATIVE FLOORING
(Bloomvont, Park Way)
(0635) 46900

THURSDAY, MARCH 3, 1994 Price 30p

Impton **KINGSCLERE** Woolton Hill Burghclere Highclere **HUNGERFORD** Kinbury Inkpen Great Bedwyn **LAMBOURN** Ramsbury Aldbourne Baydon

LATE SUMMER START FOR £77m. BY-PASS

REPORT BY
STEPHEN ROUSE

NEWBURY by-pass will be under way by

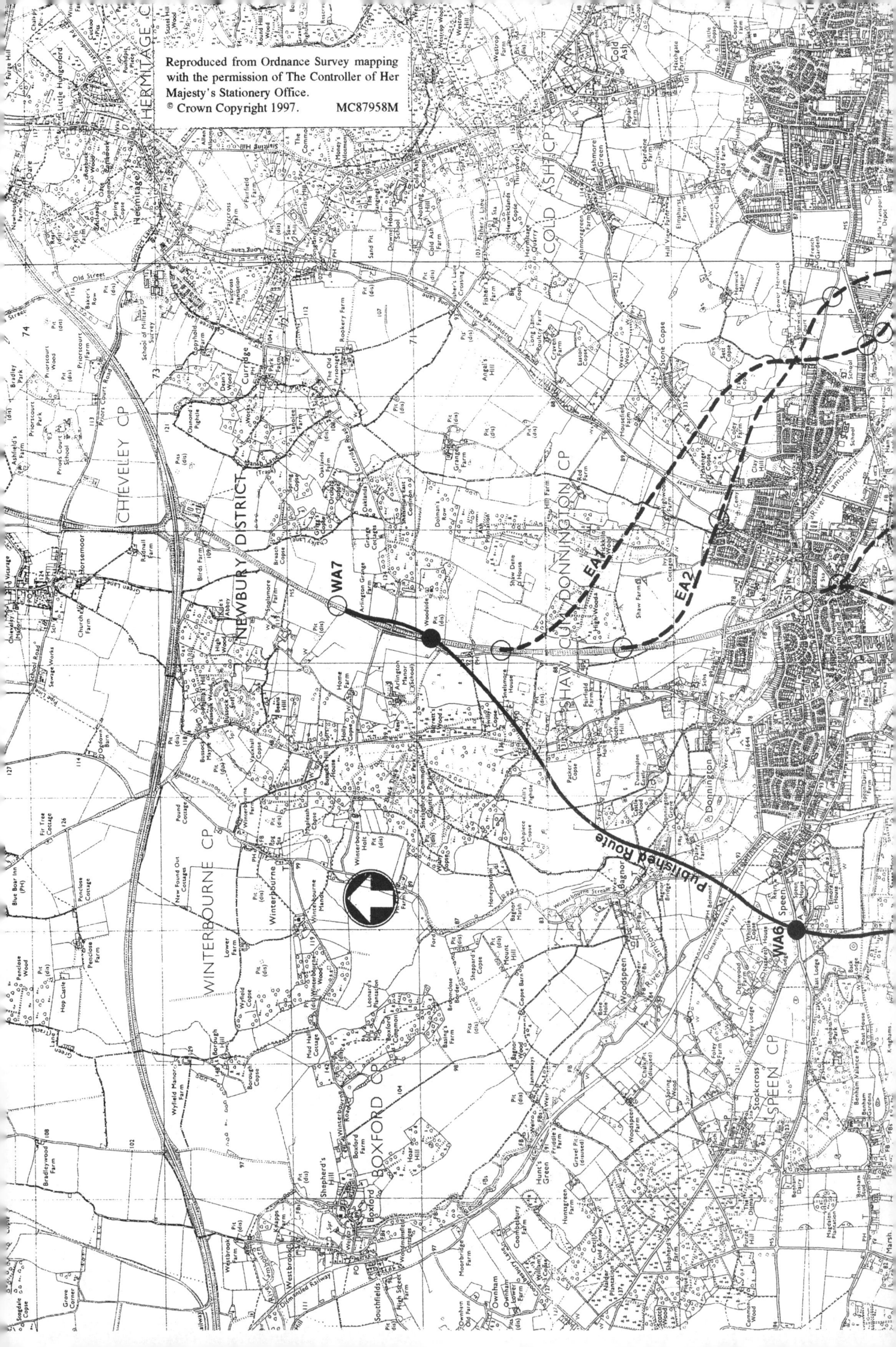

EA3

Race Course

GREENHAM CP

EA2

EA1

CA1428

CA NEWTOWN CP

CAN

CA2478
EA123

Burghclere

WA1

WA5

WA3

WA2

WA8

WA10

ENBORNE CP

Woolton Hill

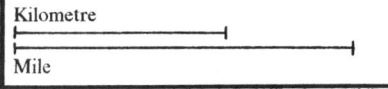

SCALE

Kilometre

Mile

The protesters mounted a concerted campaign of letter writing to the Newbury Weekly News. Emotive phrases became the norm. Twyford Down had been described by a leading Winchester protester as, *"possibly the greatest single act of visible destruction ever worked on the scenery of southern England"*. Newbury was to attract even greater hyperbole; *"The Newbury bypass is acknowledged to be the most environmentally damaging road scheme in the country, worse even than Twyford Down."* Jonathan Porritt, a founder member of Friends of the Earth, (and adviser to the Prince of Wales) in his column in the Daily Telegraph, railed against the *"transparently rigged public inquiry system"* and against the Roads Minister, the jowled Robert Key, describing him several times as *"Mr Toad"*. The "Third Battle of Newbury" protest group was formed, initially by those living near the western route (who were apparently unaware that the local museum listed the Third Battle of Newbury as already having taken place on 9 November 1644). The group attempted to whip up opposition to the bypass by claiming that the road would damage the sites of the civil war battles which took place in September 1643 and October 1644. They held a rally on 22 May 1994 at the Watermill Theatre at Bagnor, the village most affected by the bypass. The group let it be known that it intended to back an appeal to the European Commission on the grounds that the requirements of the European Environmental Impact Assessment Directive had not been met, and said it would be approaching the recently elected Conservative MEP for the area, Mr Graham Mather. It also called for a fresh inquiry under Section 302 of the Highways Act, a request with no possibility of success in view of the comprehensive nature and unequivocal conclusions of the two inquiries which had already been held. The group also claimed that the Landscape Advisory Committee's Report, which commented adversely on the western route, had not been made public. In fact the document, with plans, had been available at the 1988 public inquiry and listed in the Inspector's report as Document D56.

The Friends of the Earth attended the Watermill Rally and its members began to take an active and leading role in the protests. They condemned the effect of the bypass on the Snelsmore Common SSSI and organised another rally there in July. Their concerns were not shared by English Nature who had the responsibility of managing the Common for Newbury District Council and who, as the Nature Conservancy Council, had successfully persuaded the DoT to re-align the bypass so as to reduce its effect upon the SSSI. The area of land being acquired by the DoT was quoted as only 1.2% of the total Common area and, following some determined negotiation, three areas of exchange land, *"equally advantageous"*, in the legal phraseology, and adjacent to the Common, were to be bought by the DoT and handed over to Newbury District Council in compensation.

The impending designation of the Rivers Kennet and Lambourn as SSSIs was used by Friends of the Earth as a reason for reopening the public inquiry since they were crossed by the bypass. However, English Nature again took a more relaxed view of the situation, *"The people building the road will have to liaise with us to protect the watercourses as much as possible but we would expect that anyway."*

In August 1994, the DoT's Highways Agency, who had now taken on responsibility for road schemes, conceded that its contractual procedures for the road construction were running late. The tender documents had not been sent out until June and were not due back until September. November was being quoted as the earliest start date.

A constant theme in protesters' letters to the Newbury Weekly News was that environmental matters had not been investigated in sufficient detail. These appeared to be from correspondents who had not attended the inquiries and who had clearly not studied the bypass witnesses' proofs of evidence nor even the Inspectors' reports. No attempt was made to refute these allegations by the Highways Agency. In the light of following events, this was a very regrettable omission. By not challenging them, the misleading and often-repeated claims acquired an undeserved credence. The Third Battle of Newbury's interpretation of this silence was reported by the protesters under the headline, *"Slowly, the DoT is starting to listen"*. To engage consultants, consult legal opinion, hire halls and print and distribute literature, the Friends of the Earth with its nation-wide membership, obviously possessed resources far in excess of the local Bypass Supporters' Association, who were the only organisation refuting the protesters' propaganda at the time. The media found the simple story line of a Government trying to bulldoze a road through beautiful countryside far more newsworthy than the supporters' contention that the Government had done all it could, and that all of the other solutions were so much worse.

The Royal Commission on Environmental Pollution produced a report recommending huge increases to fuel prices, road pricing within towns and a cut in the roads programme. The report was described by the AA as being, *"characterised by breathtaking naïvety"* on the grounds that large petrol price increases had not been effective in the past, that road pricing was immensely complex to implement and that the roads programme contained no new motorways but only improvements and bypasses which reduce pollution by eliminating static queues of vehicles in towns.

One sad result of the publicity emerged. In the laudable attempts to encourage children to care for the environment, it was comparatively easy for an anti-bypass theme to be preached by those having sympathy with the protesters. The Newbury Bypass became a popular subject for school essays or larger projects. The considerable volume of information produced by the protesters' organisations was more readily available than that from the supporters. Government sources were silent on the subject. From the information received by the Supporters' Association it appeared that a bias was setting in, and it compiled and distributed a two-page summary of the factual history of the bypass to 70 schools in the area.

Battlefields

In September 1994, English Heritage sought residents' views and comments on a battlefield register which it was compiling. One of the battlefields on the register was the first civil war Battle of Newbury. The Friends of the Earth claimed that, *"The A34 Newbury bypass is set to run through the 1st Battle of Newbury Civil War site of 1643, and will significantly damage the site."* A local protester, writing to the local paper stated that, *"A battlefield is more than just the area of combat, and includes approach routes, marshalling and manoeuvring areas, sites of reserve troops, baggage trains, etc."* The huge swathe of Britain which this definition of its myriad battlefields would encompass, was not that used by English Heritage, who defined a battlefield as, *"the outer reasonable limit to the area within which the bulk of the fighting took place."* By this definition, it can be argued that the bypass runs through neither the first nor the second civil war battlefields of Newbury. The site of the second battle is largely built upon. The claim that the road runs sufficiently close to disturb the ambience of the sites is highly subjective. In any event, as English Heritage emphasises, *"It is not proposed that the Battlefields Register will have any direct legal effect, nor will inclusion of any land on the Register carry with it any legal obligations."*

The FoE Bypass

The Friends of the Earth themselves clearly believed that a bypass solution was necessary because they called a public meeting on 7 October 1994 to launch their alternative central route, the On-Line Alternative, designed for them by Steer Davies Gleave, transport consultants. Instead of a complete 8 mile road, this route only covered a one mile stretch within Newbury town centre. The design proposed huge *"traffic in the sky"* structures similar to those considered in detail at the 1988 inquiry, which would clearly have invited the same damning comments by the Inspector in respect of their dominance in the town, their damage to the *"human environment"*, the near impossibility of construction and sheer cost. In addition, the countryside damage that the route would have caused within Hampshire, was not covered by the proposal. The Friends of the Earth had labelled the western route as, *"hugely expensive"* yet their own proposal, like the other similar central routes, would have cost almost twice that of the western bypass. The Bypass Supporters' Association concluded that it could not be taken seriously. But a constant theme in the protesters' publicity was to be - why not pause and have a fresh look at the whole scheme again? This was regarded by supporters as a fall back position for protesters; if the bypass could not be stopped, then they would try to delay it as long as possible. It became clear later that the theme had struck a chord with the Transport Secretary of State.

Infilling

An often repeated claim was that bypasses invariably attracted infilling development. The Supporters pointed out in 1994 that the facts did not support the assertion. To the south of Newbury, the A34 Kingsworthy bypass at Winchester had attracted no such development in its 25-year life. Neither had it happened a few miles further north at Whitchurch. To the north of Newbury, the A34 western bypass at Abingdon could be said to have restricted rather than encouraged development over the previous 18 years, and at Oxford there had been hardly any development between the town and its 29-year-old five-mile A34 western bypass. At Newbury, for three miles of its length, the embankments and cuttings of the western bypass route follow exactly the embankments and cuttings of the old railway line which, for over one hundred years, have attracted no infilling development. The planning authorities may or may not decide to use a bypass or some other physical feature to define the limit of development. The Inspector at the 1988 inquiry had stated, *"The pressure for infilling and development is obviously a widely held concern and has been used as a reason to support objections to the PR . . . .However, if there is specifically no provision for such development in the Structure Plan, and that appears to be the case, and the local planning authority, in this case the Newbury District Council, has publicly stated that it is not accepted. . . . then I can only conclude that it must be left to the local planning authority, together with the Secretary of State, to regulate development in accordance with normal planning criteria."*

Since 1994, there has been considerable pressure on Newbury District Council to provide more housing and, in order to provide its allocated share, development has occurred to the east of Newbury, within Thatcham. Additional housing is likely to take place to the south of Newbury at first, but the county boundary with Hampshire will put a limit on the number which can be provided. Some development to the west therefore appears inevitable, irrespective of the construction of the western bypass.

The Tunnel

As with many bypass schemes, a tunnel alternative had been considered at the 1988 inquiry. Now a new tunnel was proposed by protesters. In December 1994, a large advertisement was placed in the Newbury Weekly News proposing a two-mile tunnel under Newbury with access roads through the environmentally sensitive Enborne Valley and Horris Hill areas to the south. The tunnel (actually two tunnels, one in each direction) would have been twice as long as the Dartford Tunnel and its capital cost would be more than six times that of the bypass. It would require ventilating outlets which would discharge concentrated pollution in a built-up area. Perpetual annual costs would be incurred for ventilating, cleaning and lighting. At Newbury, the recovery of these running costs and amortisation of the capital cost from commensurate tolls would be virtually impossible since traffic could readily avoid payment by continuing to use the existing A34 through the town, particularly at off-peak periods. Junctions in such a tunnel could not be provided due to increased possibility of collisions, particularly those resulting in fire with consequent risk of asphyxiation to those locked in a traffic queue. On traffic, economic and safety grounds the project would fail. The public inquiry procedures and extended construction period would put the completion of a tunnel fifteen years or more into the future and the project was clearly not a serious solution to Newbury's traffic problems.

Rare wildlife

Considerable scepticism was expressed by Newbury residents at the protesters' attempts to delay construction by the discovery of rare wildlife along the route. Badgers were reported in Penwood and elsewhere. They have statutory protection under the Protection of Badgers Act 1992 which makes it an offence deliberately to kill, injure or take a badger or to damage, destroy or obstruct a badger sett unless a licence is obtained from English Nature. The DoT had consulted English Nature and three artificial badger setts had been provided for those animals displaced by the bypass construction. Under the road, badger tunnels were to be provided at secret locations, pre-scented to encourage their use, with badger-proof fencing sited to guide the animals to them. Some nightjars were reported in Snelsmore

Common, but they occur elsewhere in the area, notably on Newtown Common where they persist in spite of the adjacent A34 traffic. A few dormice were discovered in Reddings Copse. They are a protected species, but English Nature employed contractors to capture and remove them to a sanctuary from where they could eventually be relocated in a suitable woodland habitat. In co-operation with English Nature, all major trees were surveyed for the presence of bats. A few pipistrelle and noctule bats were caught and released into adjacent woodland after removing their roosts. Some 200 bat boxes are provided in the new locations to give additional habitats. An existing brick-lined tunnel under the old railway line has been renovated and extended to allow bats to use it as a roost.

Protesters' Actions

Contracts for widening and strengthening approach roads for contractors' vehicles were placed by the Highways Agency in 1994 and there was some interference by protesters to the machinery being used - a foretaste of what was to come. The Bracknell offices of John Mowlem Construction were invaded by protesters although the company said that it had not been awarded the contract. Camps were set up by protesters along the route of the road causing bitterness among local residents who considered that the police were lax in not taking the same action to evict the protesters in the same way that they believed locals would have been removed if they had set up similar illegal camps. The direct action protest group, Road Alert, were reported to have moved to Newbury towards the end of 1994.

In addition to the publicity on environmental issues, the protesters were reported to have bombarded the Transport Secretary both with letters and with a petition of 6,000 nation-wide signatures opposing the bypass. Court action was also pursued. In the name of Miss Helen Anscomb, an attempt was made to force a judicial review of the route, early in December 1994. The judge in the case, Mr Justice Jowitt, said that she had failed to show that she had an arguable case and that, in any case, the Minister was answerable to Parliament and not to the Courts. Miss Anscomb subsequently was reported to have said that the case was likely to have cost around £14,000 and that she would have difficulty on her teacher's salary in finding that sum, but the cost was assumed to be met by some other source as no subsequent difficulty was reported.

Government Responses and Suspension of the Bypass

Rumours abounded that the bypass was likely to be delayed or cancelled. In an attempt to counter the protesters' propaganda, the Bypass Supporters' Association had prepared a 15-page 'Facts File' summarising the history of the bypass and had distributed copies to the media as well as to interested organisations and individuals. A copy was sent to John Watts, the Minister for Railways and Roads. In a letter to the Association dated 31 October 1994 he replied;

"Thank you for sending me a copy of your 'Facts File' on the A34 Newbury Bypass. I have also received a copy of your leaflet 'Response to the Central (FoE) Alternative 'On-Line' Proposal and other Current Issues. I am most grateful to your Association for its efforts in demonstrating support for the bypass proposals. I hope you will continue to do so.

I am quite satisfied that the bypass is urgently needed - indeed, I have experienced for myself the awful congestion in Newbury and sympathise with local residents who have waited so long for relief from delays, pollution and not least for a safer road through the centre of the town.

I assure your Association that against the background of public consultation, two public inquiries and a period for High Court Challenge, I can see no grounds for further delaying this much needed scheme.

You will be pleased to know that tenders for construction of the bypass have been received and are being assessed. We expect to appoint a contractor shortly for a start of work as soon as possible.

Best wishes for your continuing campaign. "

The Leader of Newbury District Council, Councillor Keith Lock, received a letter dated 1 November 1994 from the Chief Executive of the Highways Agency with very similar wording to John

Watts' letter to the Supporters' Association. However, the rumours persisted. Later in November, road cuts were announced in the Budget. At the beginning of December, the Highways Agency said that it was extending the period for consideration of tenders for the bypass. On Wednesday, 14 December, the Supporters' Association telephoned the office of the Minister to enquire whether the assurances given in his letter of 31 October were still valid. After internal enquiries, they were telephoned back to be told that there had been no change.

But the adverse publicity was clearly having an effect on the Secretary of State for Transport, Dr Mawhinney. On Monday, 19 December 1994, he stood up in Parliament to announce a *"re-focusing"* of the Roads Programme. Saying that the trunk road network was broadly complete, he went on to list improvement schemes on the M1, M25 and in London and elsewhere. He then dropped his bombshell by announcing,

"In the case of one other scheme, the A34 Newbury Bypass, I have concluded that work should be delayed whilst further consideration is given to the proposed route. I reached this view after a private visit to Newbury to examine both the existing A34 and the proposed bypass route. I have asked the Highways Agency to look again at the plans for the bypass and to explore other options. Once the Agency has reported back to me I will make a further announcement. I would expect to be in a position to do this in about a year's time, when I announce road scheme starts for 1996-97."

This double-fatality accident at the northern end of the Sandleford Link in October 1995 is one result of a dangerous mix of cars and heavy vehicles ©PB

CHAPTER 7

Reaction of the Local Community

The gleeful protesters greeted the decision as a triumph. Within Newbury, there was consternation. The Bypass Supporters' Association were quoted in the headlines as saying, "THE GOVERNMENT HAS DECEIVED US". The Newbury MP called the decision *"appalling"* and said the town had been *"betrayed again"* and that *"The baby has been thrown out with the bath water"*. In a Radio Five phone-in programme, the Minister for Railways and Roads was accused of duplicity on Wednesday, 14 December 1994 in giving his assurance of a go-ahead, but he replied that he himself had not known of Dr Mawhinney's decision until the evening of Friday, 16 December, less than three days before the announcement, a statement which, at the very least, seemed to indicate a flaw in the communication and co-ordination systems within the DoT. The chief executive of Newbury District Council said, *"We have been inundated by calls from the public incensed at the Minister's decision. They are looking to the Council to take a lead in the battle for the bypass."*

Dr Mawhinney refused to give details of his Newbury visit to journalists but he was required to give an answer to a Parliamentary question by the Newbury MP, in which he said that he paid a three-hour visit on 28 November, driving from north to south along the A34 and visited a number of sites to the west of the town. *"I got mud on my shoes and I also got held up on the A34"*, he said. Apparently he was accompanied only by his private secretary and the chief executive of the Highways Agency, who admitted that £900,000 had already been spent on bypass preparatory work. However Dr Mawhinney met neither the protesters, supporters, local council members or officials, police, residents or any member of the general public. There was general astonishment that the Secretary of State had the power to suspend the results of fifteen years of such intensive, democratic investigation and call for further studies on the basis, as it seemed, of a personal whim. The phrase, *"in about a year's time"* was regarded as prevarication, allowing a slippage of appreciably more than 12 months; perhaps disguising an abandonment of the scheme altogether. The statement that *"other options"* were to be explored seemed to indicate that a new route, or a major modification, would be investigated and recommended, inevitably leading to a new public inquiry and a delay of an absolute minimum of five years with eight or nine years being more probable if the timescale to date was to be re-enacted. Even if the Highways Agency's new report eventually concluded that there should be no change, the work would certainly have to be put out for re-tendering, a process which would add an additional four months or more to the total delay.

The measures to solve Newbury's increasing local traffic problems would be pushed well into the future. The new Sainsbury's store had been given planning permission by a majority of council members, with some reservations, on the assumption that the bypass route had been settled and would soon be built. It was opened in 1994 and, with its 600-space car park on a busy corner of the A34 in the town centre, it was already adding appreciably to local traffic.

The Highways Agency team to look into the bypass issues was to be composed of personnel who had no previous involvement with the bypass. (Later it emerged that they were taken from the North West Construction Programme Division based in Manchester.) The implication was that the existing staff at Dorking were not to be trusted to produce a unbiased conclusion. Speculation was rife about Dr Mawhinney's motives. He had taken office as Secretary of State for Transport only the previous summer and it was his predecessor, John MacGregor, who had sanctioned the bypass. Asked if the decision was a punishment for Newbury having elected a Liberal Democrat MP, he replied, *"Those people saying that don't know me very well."* Other motives were considered. The suspicion grew that he had become alarmed by the thousands of letters he had received from protesters and by the anti-bypass media coverage. The protesters had promised demonstrations larger than those at Twyford Down during the period of construction, which would have taken place during the run-up to the General Election. From the Government's viewpoint, there was political advantage in delaying the start of the scheme until after the latest date for a General Election in mid-1997. The delay would also save costs at a time of budget stringency. If the bypass was to be restored in the roads programme, Dr Mawhinney would need to be persuaded that there was greater electoral harm in delay than by building the bypass quickly. He appeared to have chosen the timing of his announcement with care; coming immediately before the Christmas recess it could have been expected that opposition to it would be muted due to the holiday period. However the residents' representatives wasted no time in organising themselves.

Formation of the Newbury Bypass Forum

The Newbury District Council decided to call a meeting of representatives of local interests with the intention of forming the "Newbury Bypass Forum". The inaugural meeting was held on 4 January 1995. Those present included three Councillors from Newbury District Council, a Councillor from Basingstoke and Deane Borough Council, officials from Berkshire and Hampshire County Councils and representatives of Berkshire Fire and Rescue Service, Thames Valley Chamber of Commerce, the Kennet Shopping Centre, Newbury Buses and the TGWU, Newbury Racecourse, Newbury Bypass Supporters' Association and local solicitors as well as Vodafone and Bayer, the two largest employers in the town. Graham Mather MEP and the Earl of Carnarvon were also present, the latter being the Chairman of the South East Regional Planning Board. David Rendel, the Newbury LibDem MP, took the chair. An apology was received from Sir David Mitchell, the Conservative MP for NW Hampshire but he agreed to act as alternate chairman with David Rendel, a decision which gave emphasis to the apolitical nature of the Forum. Although some support for the bypass was expressed by members of the local Labour Party, it seemed that internal differences of opinion prevented them from accepting an invitation to join the Bypass Forum. Richard Benyon, a Newbury Councillor and prospective parliamentary Conservative candidate, was appointed treasurer. The Forum clearly represented a wide cross-section of local interests and, later, was to be accepted by the Highways Agency as their link with the community in the Newbury area. The Forum passed resolutions to raise funds for a locally driven "Bypass Now" publicity campaign and to form a Parliamentary panel and an action committee.

The Restoration Campaign

A petition for the restoration of the bypass had already been initiated by the Newbury MP. This was now taken over by the Forum. Postcards were distributed for people to sign and send to Dr Mawhinney. Car stickers were printed - NEWBURY BYPASS NOW! - which local shopkeepers, supermarkets and petrol stations proved only too willing to give to customers. Leaflets setting out the case for the Western Bypass were printed, explaining the pollution being caused by the existing road, the environmental and traffic advantages of the western route over any alternative, and detailing the history of the bypass to date and the decisions which had emerged from the public inquiries.

Both MPs were vigorous in pressing the case for a bypass in an adjournment debate successfully sought by Sir David Mitchell in late January. An indiscreet letter from the Highways Agency to Newbury District Council inflamed the situation; *"Even if minor changes are made to the route, new legal orders may be needed. These could take some years to go through the statutory processes, so for the purpose of development control, I recommend that we now work on the basis that there will be no additional capacity through Newbury before 2003."* The Highways Agency chief executive later apologised for the *"misleading messages"* given in the letter.

On 24 January 1995, Lord Carnarvon put questions in the House of Lords to Viscount Goschen, the Parliamentary Under-Secretary of State for Transport, but described the answers he was given as *"innocuous"*. On 6 February, the Bypass Forum held a packed public meeting chaired by Newbury's MP at St Bartholomew's School in the town. A few prominent protesters attempted to shout down those speaking in favour of the bypass but the strength of local feeling was made evident by the public reaction against them.

Dr Mawhinney agreed to meet the two MPs on 15 February and said that each could be accompanied by two constituents. No doubt to his surprise, he had been lobbied over his bypass decision at a meeting of the Confederation for British Industry earlier in the day. The chairman of the South East region of the CBI had said, *"It's true we don't take to the streets; it's true we haven't recently chained ourselves to bulldozers, but we know how much of our business costs are transport costs and how vital it is that we achieve an effective, basic network in the shortest possible time . . . the abandonment of the Newbury Bypass on the eve of work commencing is a worrying sign of the way the Government is thinking."* At the meeting with the MPs, relaxed on a low seat in his customary shirt-sleeves, Dr Mawhinney listened to questions from those present. The Bypass Supporters' representative said that the question that everybody in Newbury wanted to ask was why he considered that fifteen years of careful, continuous, democratic investigation had not already produced the optimum solution to Newbury's main traffic problem, but he refused to reply. The offer was then made to help speed up the deliberations of the new team, who were obviously strangers to the area, by supplying purely factual information about the locality and the traffic problems but he declined the offer, presumably not wishing the team to be influenced by local contacts. The Newbury MP asked what other options were being investigated but Dr Mawhinney said that he had difficulty replying to that question because of legal advice. It was assumed that his concern was that the disclosure of a particular potential alternative might affect property values along its route, but what had apparently been overlooked by him and his advisers was that procrastinating on the western route would also have an effect on property values and cause planning blight throughout the whole area within and around Newbury where possible alternatives might now be considered. Any traffic plans for west Berkshire and north Hampshire had been thrown into the melting pot.

The extraordinary decision to shelve the bypass was given prominence by the media. For Meridian TV, the bypass was the principal topic of news for its viewers in the Newbury area, where one of its three studios is located. There was an item on their local news bulletin on most evenings. The media usually turned to the Lib Dem Newbury MP, David Rendel, to put the case for the supporters, and he worked tirelessly on behalf of the Bypass Forum in the collecting of information, drafting of press releases and other matters. The involvement of Sir David Mitchell MP, Graham Mather MEP and Richard Benyon, all Conservatives, was undoubtedly of major importance in demonstrating that the campaign was not politically motivated but truly represented the will of the local community. Later, Clare Short, then the Labour shadow Minister of Transport, was to write to the Bypass Supporters' Association saying that, although the Labour Party were committed to examining all new road-building proposals on a case-by-case basis, *"It will not be possible to apply this overarching review to the Newbury western bypass since the scheme now has the go-ahead."* Nevertheless a view persisted that a Labour government would be more likely to back the protesters' view of the bypass and that putting off a decision until after a General Election ran the risk of at least further delay to the project.

Prompted by the considerable public interest, Meridian TV broadcast a 30-minute studio discussion programme on 16 February. A similar programme was broadcast nationally a short time later by Central TV, hosted by "Kilroy". Dr Mawhinney was challenged by the Chairman of Newtown Parish Council to set out his reasons for the delay on a phone-in programme on BBC2, but only said that he recognised the problem at Newbury and an announcement would be made eventually. On 24th March, Dr Mawhinney attended the AGM of the NW Hampshire Conservative Association as the guest speaker. No doubt to his surprise, he was once again to be lobbied on the need for a bypass at Newbury, this time because of the danger of maintaining a single carriageway road and delaying the relief needed to the Newtown Common area within Hampshire.

On 14 March, in a debate in the House of Lords, six peers questioned Lord Goschen about the delay on the bypass, but received the stock reply that review was necessary in the public interest to ensure that the correct solution was chosen. During March of 1995 the Bypass Forum set up an exhibition in an empty shop in the Kennet Shopping Centre in Newbury. The Highways Agency had loaned the three-dimensional model of the bypass which had been used at the public inquiries together with the coloured exhibition panels showing the plans and longitudinal sections of the route to a scale of 1:2500. Information leaflets had been prepared by the Forum on topics such as the history of the bypass since 1979, the reasons for a bypass being needed, the disadvantages of the central and eastern alternatives to the western route, and details of the increasing pollution levels in the town which had been measured by Newbury District Council. Callers were invited to sign a petition calling for work to start on the bypass. Printed postcards were available to be filled in with the personal reasons for wanting the bypass. These were collected and sent to Dr Mawhinney's office via the MPs. On 28 March, the Newbury and NW Hampshire MPs, attended by supporters, handed the 10,500 signature petition to the Roads Minister at his office in Marsham Street in London.

During this period, when the arguments for and against the bypass were being resurrected, the DoT and their Highways Agency remained silent and the burden of refuting the protesters' continuing and misleading claims fell upon the Bypass Forum members. The DoT asked for the return of their bypass model; it was clearly their intention to remain neutral in the debate.

Berkshire County Council and Hampshire County Council's Environment Committee passed formal resolutions wholeheartedly condemning Dr Mawhinney's decision and supporting moves to restore the bypass.

Pressures mounted. Accidents on the A34 near Newbury were reported by the media and prompted renewed public demands for a bypass. Early in April a van driver was killed on the Newtown Straight. Two weeks later, a caravan was destroyed in a collision with a lorry at the same location. A car overturned on the A34 north of Newbury in the same period. A commercial property firm reported that businesses were being deterred from locating in Newbury because of the traffic problems. Newbury Racecourse announced that it would not go ahead with a £4 million improvement programme until the bypass was started. A major employer, Bayer UK, said that it would not renew planning permission to build an extension to its offices in Newbury because of traffic problems.

An important element in the restoration campaign was to gauge the depth of public opinion on the issue. Claims had been made by the protesters that the public had became more opposed to the road since the 1988 Inspector's finding of *"overwhelming public support"* for it. Large and small surveys indicated otherwise. In December 1994, within days of Dr Mawhinney's decision to shelve the scheme, the Newbury Weekly News conducted a seven-day phone-in poll of its readers. Of the 2,341 people who voted, only 340 supported the Minister's decision, a 6:1 majority favouring the work going ahead immediately. In January 1995, a Newbury College student carried out interviews of shoppers in the Kennet Centre in Newbury and found 6:1 in favour.

Early in February, Meridian TV conducted a phone-in poll of its own, clearly covering a very wide area. The result was announced on 15 February and again showed a 6:1 majority in favour.

These consistent figures had been drawn to Dr Mawhinney's attention at the meeting with him later on 15 February. Early in May, a number of protesters, including Green Party candidates, stood in the local elections. Without exception, they were all very heavily defeated. In the Falkland Ward, Helen Anscomb received only 3.5% of the votes. In the Speen Ward, Jill Eisele, a dedicated "Third Battle of Newbury" protester, received 2.8% and in Hungerford, Susan Millington, the active Friends of the Earth representative, received a mere 1.6%. In Craven Ward, the two Green Party candidates combined received 5.5%. The overall evidence was clear; the vast majority of the Newbury and District electorate wanted the bypass to be built quickly.

Signs emerged that the campaign was having an effect. Early in June 1995, in answer to a parliamentary question from the Newbury MP, the Roads Minister said that the Highways Agency's review team had made *"substantial progress"* and that *"the Secretary of State has received preliminary findings on certain aspects of the review"*.

The Restoration Announcement

On 5 July 1995, Dr Mawhinney left his post as Secretary of State for Transport to take over the chairmanship of the Conservative Party and responsibility for their forthcoming General Election campaign. (He was to be given a knighthood in John Major's Retirement Honours List.) His last act was to announce the conclusions of his review team and he went on to say;

"I have carefully considered all the changes and other factors addressed in the team's report, the scope for alternative options to those examined at the Public Inquiry, and the balance between all the options. I have also taken into account all the advice I have been given and representations I have received since this study was inaugurated.

I agree with the team's assessments, and with their overall conclusion that the western bypass, for which Statutory Orders have been made and confirmed, is the most effective solution for Newbury. I now intend to move ahead speedily with the proposed scheme to bring much needed relief to the town and its residents. On the basis of the team's report and the work leading up to the earlier decision, I can see no justification for further investigation or delay."

The decision was widely welcomed, although there was caution among some supporters. They had learned that the Secretary of State had the power to reverse decisions instantaneously on a personal whim, as it seemed. Dr Mawhinney had reversed the firm decision of his predecessor and Dr Mawhinney's successor was Sir George Young, dubbed the 'cycling baronet' and reputedly a member of the Friends of the Earth. These fears were soon allayed by firm statements from the new Secretary who made it clear that he intended to proceed with the bypass.

The Review Team's Report

The terms of reference of Dr Mawhinney's review team were quoted in their report; *"The aim of the study will be to look again at the published route and any other practical alternative options for reducing congestion at Newbury including, but not limited to, those already considered."*

The team considered the changes which had occurred since the 1988 inquiry. They said that the construction of a new hotel, The Hilton, and of the new Sainsbury's store formed a significant constraint on a central route. They discounted fears of the effect of a western bypass on urban development to the west of Newbury, saying, as the 1988 Inspector had done, that this was a matter for the planning authorities and that, *"There is no evidence to suggest that the planning authorities have weakened in their resolve to oppose infill development."*

They noted the changes which had occurred in environmental appraisal, particularly the publication in 1993 (with more recent updating) of 'The Design Manual for Roads and Bridges', (DMRB) Volume 11 of which *"provides the practical manual of application of all the philosophy, science, debate and legislation of the past years"*.

The presence of badgers, dormice and bats were noted, but the team said, *"Appropriate action, meeting the requirements of relevant legislation and of English Nature, has been taken to relocate them."* The team commented that no information existed on the likelihood of protected species being found on other routes. The designation of the Rivers Lambourn and Kennet as SSSIs was noted as being primarily to conserve the interest of the rivers themselves and was being done, in the words of English Nature, *"in the full knowledge that the Newbury Bypass is likely to be constructed through the area of search."*

On the battlefields issue, the report pointed out that, *"The western bypass, like the disused railway which preceded it, is on the fringes of the battlefield"*, and that, *"We conclude that the registration of the battlefield does not materially affect the scale of the impact of the western bypass or of the balance of impact between the western and other routes."*

Since the 1988 inquiry, new National Road Traffic Forecasts had been released. These indicated that the traffic flows on the Inner Relief Road within Newbury would be reduced by more than 60% by the year 2010 if the bypass was built. The opening of the Channel Tunnel had not produced any evidence of transfer of traffic from the A34 through and beyond Newbury. (As forecast by the Bypass Supporters' Association in their evidence to the 1988 Inquiry).

A report by the Standing Advisory Committee on Trunk Road Assessment (SACTRA) had predicted that new roads tended to induce more traffic. The review team's conclusion was that, *"on the basis of local circumstances . . there may be limited scope for induced traffic and . . assuming the worst case . . there could be an additional 10% of traffic on the bypass."*

The team reported that a total of £5,803,000 had been spent or committed on works and property and that the expenditure on works (£2,444,000) would be forfeited if the western bypass were to be abandoned or replaced by an alternative scheme. The team's opinion was that, if there was a switch to a different option to the western bypass, then a delay, *"could easily be ten years, with no great confidence of success."*

No doubt in response to the protesters' claims, including those of Transport 2000, the team considered the various possibilities of transferring traffic to rail, traffic management measures and a "do minimum" improvement of the existing road. Their conclusion was; *"There is no scope for a different approach, without major road construction, to the problems of congestion in Newbury. The bypass would provide fully for future trunk road needs within the A34 corridor replacing the existing road through the town. It would not, of itself, provide a complete answer to traffic congestion in Newbury but it would give an opportunity for a new approach to traffic and transport management in the town."*

The review team re-examined the reasons for the Inspector's unequivocal recommendation of the western route. They believed that, in turning down the central routes, the arguments on air quality were overstated, (exhaust emissions had improved) whereas the effects on the movements of pedestrians and cyclists and upon community severance were, *"very much under estimated."* They went on to say, *"there is no central route which can meet all of the Inspector's criticisms, and attempts to meet one particular criticism only aggravate another."*

The team went on to examine a new eastern route; it appeared obvious that they had been asked to do so. The closure of Greenham Common air base had provided an opportunity to route an eastern bypass across the old airfield. It could serve both the A339 traffic from Basingstoke although, *"contrary to what might be expected the eastern route does not offer a great advantage in respect of access to major commercial/industrial areas east of the centre,"* i.e., those at Thatcham. It also would be unable to serve traffic from *"the A343 Andover Road nor the A4 (west) traffic wishing to travel to or from the south."* Environmentally, the new eastern route would not avoid the destructive crossing of the Thatcham Moors Nature Reserve. It would require a new major junction replacing the Swan roundabout and, to the south of the Air Base, it would cut, *"centrally through the Greenham and Crookham Common SSSI (also ancient woodland)."* North of the Air Base, it would intrude into either the Newbury and Crookham Golf Course or the Bowdown and Chamberhouse Woods SSSI, which was also a nature reserve. The team did not even consider the environmental disadvantages of the section through the populated areas of Newtown and Burghclere to the south; there was clearly substantial evidence against that route without a separate investigation of the landscape and ecology problems that would arise within the Hampshire section. It reported, *"Given these additional impacts, with no significant environmental savings in compensation, it must follow that the new route is at a considerable disadvantage in respect of total environmental impact when compared with the western bypass."*

The team's overall conclusion was, *"We are satisfied that the western bypass is the best scheme for a Newbury Bypass and that there have been no changed circumstances which collectively point to a different conclusion. We are also satisfied that the made line for the bypass is the correct one."*

Those supporters who had followed the various detailed arguments at the public inquiries had expected no other conclusion and they were still exasperated that Dr Mawhinney had found it necessary to ask a team to investigate what seemed to them a cut-and-dried issue. However, the report did help to refute many of the claims being made by the protesters, in particular that the 1988 recommendations were out of date. The report made it clear that the case for a western route was as strong as ever and that it was the least damaging environmentally.

'OUTSIDERS' CRITICISED FOR PROTESTS

EURO THREAT TO BY-PASS

The protesters were furious with the decision to restore the bypass, calling it a betrayal. Friends of the Earth described it as, *"a serious breach of faith"*. Dr Mawhinney had previously said that he expected to make an announcement, *"in about a year's time"*, i.e. in December 1995, one year after his shelving of the scheme. Protesters now pinned their hopes on an appeal to the European Commission on the grounds that no proper Environmental Impact Assessment had been carried out.

CHAPTER 8

Environmental Impact Assessment

The concept of Environmental Impact Assessments and Statements was devised and formally established in the USA by the US National Environmental Policy Act of 1969, albeit for Federal projects only. There is no doubt that the intention of the EEC in establishing a similar uniform system throughout its member states is laudable. Problems such as atmospheric and river pollution transcend national boundaries. Proper environmental assessments for major projects are now obligatory and the results are required to be published in the form of Environmental Impact Statements. Protesters continually hold up the Netherlands as being in advance of the UK in its readiness to embrace environmental protection measures. However it is inevitable that environmental problems will assume greater priority in countries such as the Netherlands or Belgium which are far more densely populated than other European countries.

No matter what depth of environmental investigation is carried out, the claim can always be made by opponents that it is insufficient. They can also argue that any proposed alternative scheme should be investigated just as rigorously. Common-sense has to be applied. Where alternatives are likely to be firmly rejected for other reasons, as at Newbury, a costly in-depth environmental investigation of each them would be pointless.

EEC Directive 85/337/EEC - Environmental Impact Assessment

Both the Third Battle of Newbury and the Friends of the Earth claimed that an Environmental Impact Assessment was required under the terms of European Directive 85/337/EEC and that such an assessment was not carried out by the UK Government. These claims were repeated several times in letters and statements to the media and various approaches had been made by the protesters to the European Environmental Commissioner's office since 1990. The protesters' efforts were intensified following the restoration of the bypass.

There was no doubt that the Newbury Bypass, being part of a major trunk road, fell within the definitions of Annex I of the Directive, i.e. a scheme for which an Environmental Impact Assessment was mandatory. However, the basic legal argument revolved around dates. The Directive was dated 27 June 1985 and it was notified to Member States on 3 July 1985. Each Member State was required to introduce its own internal legislation to implement the Directive within three years of the notification date, i.e. by 3 July 1988. So far as road projects were concerned, the UK Government implemented the Directive by inserting a new section, 105A, into the Highways Act 1980, but it did not do so until 21 July 1988, eighteen days after the deadline. (Some other countries, notably France and Germany, did not implement the Directive until weeks or months later.) So far as the European Commission was concerned, these late implementations were to be ignored: the significant date was 3 July 1988.

The European Commission considered that projects were required to conform to the Directive if they had not been approved by that date. The UK Government disagreed, considering that projects were caught by the Directive only if they had not been **initiated** by that date. The Newbury Bypass was one of the cases 'in the pipeline' on the significant date since it had been initiated by the publication of the draft line orders on 30 October 1986 although final approval was not granted until 6 July 1990.

The Supporters' Association had heard that the European Environmental Commissioner, Carlo Ripa Di Meana, had written an official 'first letter' to the UK Government regarding the lack of an Environmental Impact Assessment (EIA). The Association protested at this action in a letter to the Commissioner dated October 1991. In February 1992, the Commissioner replied saying that the matter was being investigated. The Association received a response in August 1994 (eighteen months later !) from the Commissioner's office, signed by L Kramer, saying that, *"The Commission is of the opinion that, although no formal environmental assessment was carried out in respect of the project, the procedures followed in practice met the requirements of Directive 85/337 on Environmental Impact Assessment."* The letter appeared to be a standard one, probably sent to all those who had raised the issue.

However, the protesters continued to pursue the matter in Brussels. Where such a disagreement exists, the final decision lies, not with the Commission, but with the European Court of Justice (ECJ). Neither the Newbury Bypass nor any other UK 'pipeline' case was taken to the ECJ, but there were two relevant German cases before the Court in which the UK Government's lawyers participated, due to the UK's interest in the matter. The ECJ's decision on a particular case is frequently not published until many months after the hearing. To provide an early pointer to the Court's thinking, it is customary for the particular Advocate General to publish his opinion shortly after the hearing. It is very unusual, though not unknown, for the eventual decision of the Court to differ from its Advocate General's early opinion. In the case of 396/92, a Bavarian road scheme, the Advocate General, Gulmann, on 3 May 1994, came down in favour of the UK Government's view - that projects already initiated by 3 July 1988 were not subject to the Directive. The Court's decision did not refer to the issue.

The second project was an extension to a German power station, 431/92. Here again a different Advocate General, Elmer, in his opinion delivered on 21 February 1995, also agreed with the UK Government's position.

The Commission were obviously waiting for the final decision of the Court on this latter case to be published, to see whether it would disagree with its Advocate General's opinion. The decision was published on 11 August 1995. It did not address the question of pipeline cases; it did not have to do so since it came down against the Commission's case for other reasons. In the absence of a ruling, the opinions of the Advocates General gave the only pointer as to the likely outcome of any ECJ action on Newbury.

On 28 September 1995, a small delegation from the Newbury Bypass Forum attended a meeting in Brussels with the new Environmental Commissioner, Mrs Ritt Bjerregaard, arranged by the Newbury MEP, Graham Mather. The Newbury District Council representatives explained the route of the proposed bypass and showed photographs of the attractive, though not exceptional, countryside through which it would pass. The Supporters' Association representative explained the history of the investigation including the information which had been made available to the public on the environmental impact of the scheme. It became clear at the meeting that the only issue of interest to the Commissioner was whether or not any of the stipulations in the Directive had been breached and, if they had, whether it was worth indicting the UK Government at the ECJ. Their decision had not been made. A meeting of all the Commissioners was due to take place at the end of the month and those cases to be put before the ECJ by the EEC over the next few months were to be identified. It was obviously desirable that the Newbury Bypass should not be on the list.

There was nothing that the Forum could do over the argument on dates. There were no special factors at Newbury which could add to the evidence already presented in the two German cases. However there existed considerable information to prove that the actions taken by the UK Government at Newbury had satisfied the Directive requirements, as L Kramer of the Commissioner's office had himself believed when writing his letter in August 1994, referred to above. It would be pointless of the EEC to pursue an expensive and doubtful Newbury Bypass case at the ECJ over dates if they were going to lose the issue anyway over satisfying proof.

On returning from Brussels, the Forum members promptly compiled a list of the requirements of the articles of the Directive listing against each one the evidence at Newbury which satisfied the particular condition. The DoT had already sent its response to the Commissioner but the Forum considered that reinforcing evidence sent from a different and local source might influence the Commissioner's deliberations. This information was sent to the Environmental Commissioner via the MEP:

Articles 3 & 8 state that the assessment should identify and take into account the effects and inter-action between humans, fauna, flora, soil, water, air, climate and the landscape as well as the cultural heritage.

The DoT can no doubt claim that they took all those matters into consideration in carrying out their assessment.

Article 5(2.) requires that the following information is made available to the public and summarised in a non-technical form:

- A description of the project
- A description of measures to cope with adverse effects
- Data on the main environmental effects

The evidence for meeting these requirements is as follows:

*1.   **June 1982:** 18,000 copies of a Brochure together with a postage-paid Questionnaire were posted to households in and around Newbury. The Brochure included a detailed map showing the four main routes under consideration by the DoT to a scale of 1/25000. It gave a list of the principal environmentally-sensitive areas on each route and indicated by symbols and colour the position of Areas of Outstanding Natural Beauty, Areas of Great Landscape Value, Common Land, Recreation Areas, Conservation Areas, Sites of Special Scientific Interest, Scheduled Ancient Monuments, Nature Reserves, Archaeological Sites, Listed Buildings and other sites.*

*2.   **July 1982:** Plans and other relevant material were put on display at Public Exhibitions held in Newbury between 7 & 10 July and 16 & 17 July and in Burghclere on 13 & 14 July where staff of the Department of Transport and their consultants were present to answer questions.*

*3.   **19 June 1984:** The DoT announced their preference for a western route, with reasons, in a 5-page Minister's Statement, saying that further studies would be carried out on ecology, noise and landscaping and that the effects on people and agriculture would be further considered. The announcement was made by press notice and separate letters to interested organisations.*

*4.   **30 October 1986:** The five draft Orders describing the route of the line of the Bypass were published in the press announcing the intention to initiate the consent procedure for the western route.*

*5.   **November 1986:** Plans to a scale of 1/2500, a topographical model, display panels showing landscaping and perspective views and explanations of traffic and noise and other material were put on display at Public Exhibitions held in Newbury between 6 & 8 November, in Burghclere on 11& 12 November and in Stockcross on 14 & 15 November with staff on hand to answer questions. A Brochure showing a plan of the western route to a scale of approximately 1/15000 was made available showing similar environmental information to that in the 1982 Brochure described above.*

*6.   **14 April 1988:** Plans and descriptions of 12 alternative routes and 8 variations to the western routes from objectors to a scale of 1/25000 were published in the press by the DoT's consultants.*

*7.   **May 1988:** A pre-inquiry public meeting was held in Newbury on 11 May outlining the procedure to be followed at the public inquiry. Before the start of the inquiry, 59 documents were deposited in the inquiry library by the DoT and copies of the 58-page Statement of Reasons were made available to the public together with comprehensive papers detailing the DoT's case on the effects of Traffic, Landscaping, Noise, and Ecology including measures to cope with adverse effects.*

*8.   **14 June to 9 November 1988:** A public inquiry was held into the routing of the bypass at Stockcross, near Newbury. 3 alternative eastern routes and 7 alternative central routes including a tunnel were put forward. 141 sets of documents were presented and 74 sets of plans were scrutinised. 147 witnesses gave evidence. Environmental issues dominated the inquiry.*

*9.   **24 March 1993 to 16 June 1993:** A public inquiry into the side roads, compulsory purchase orders and exchange land certificate for Snelsmore Common was held at Stockcross. 142 sets of documents were presented and 60 witnesses gave evidence. Environmental matters were again prominent in the investigation, principally in relation to an area of National Trust land adjacent to 'The Chase'.*

The documents, exhibitions and public inquiries satisfying this Article indicate that a considerable volume of information was made publicly available.

Article 6(2.)   requires that the public be given the opportunity to express an opinion before the "project is initiated".

The public exhibitions in July 1982 and November 1986 together with the long primary public inquiry (14 June 1988 to 9 November 1988) and secondary inquiry (24 March 1992 to 16 June 1992) gave adequate opportunity for this provision to be met.

Article 9 requires the competent authority to inform the public of the reasons for, and the content of, the decision to proceed.

The main 28-page decision letter on the primary inquiry was published on 16 July 1990 and was accompanied by the 488-page Inspector's Report. The 18-page decision letter on the secondary inquiry was published on 24 March 1993 and was accompanied by the 122-page Inspector's Report. Publication of the 44-page independent Highways Agency Study Team report in July 1995 updates the reasons for the decision. These documents fulfil the obligation under Article 9.

To what extent the EEC were influenced by this evidence is not known, but just over a fortnight after it was sent, on 20 October 1995, the Commission issued a press release announcing that:-

"In the light of a decision from the European Court of Justice in August 1995 (C431/92), the Commission has decided to interpret the Environmental Impact Assessment Directive (85/337/EEC) as requiring this assessment only for projects where the procedure for consent was started after 3 July 1988, i.e. after the latest date for transposing the Directive into national law.

That means that the EIA Directive does not apply for two road projects, the Newbury Bypass, England, and the M77, Scotland. Although development consent was given for these projects after 3 July 1988, lengthy consent procedures started well before that date.

Consequently decisions to proceed with the road projects and the actual routing have to be considered a purely national matter."

The protesters fought against the decision, appealing both to the Transport Commissioner, Neil Kinnock, and to the European Ombudsman, claiming that the Commission was guilty of maladministration in saying that an EIA was not required, but the President of the European Commission, Jacques Santer, backed the Environmental Commissioner's view. The evidence was clear. The new statement and Kramer's earlier letter showed that the EEC's view was that the Newbury Bypass had breached neither the letter nor the spirit of the Environmental Impact Assessment Directive.

EEC Directive 92/43/EEC - Conservation of Natural Habitats

The UK Government has satisfied the EEC conditions required under this Directive. In the United Kingdom, the Conservation (Natural Habitats etc) Regulations 1994 extend the Wildlife and Countryside Act of 1981 in establishing a system of strict protection for the animal species listed in Annex iv (a) including dormice and bats. This legislation makes it an offence deliberately to disturb, capture or kill any protected species or to damage or destroy a breeding site or resting place. However no offence is committed if protected animals are moved for the purpose of conserving them under a licence granted by the Government's scientific adviser, English Nature. The relocation of dormice and bats on the route of the Newbury Bypass was authorised by such a licence.

EEC Directive 79/409/EEC - Conservation of Wild Birds

This requires that Member States designate Special Protection Areas for the conservation of species. None of these areas exist on the route of the bypass. The nightjar population of Snelsmore Common, only 2 or 3 pairs, is insufficient for designation as a Special Protection Area. In any event, the small area of the common affected by the bypass is unsuitable as a nightjar breeding habitat.

CHAPTER 9

Placing of Contract

Although Dr Mawhinney's announcement had come after seven months instead of the twelve months given in his "shelving" statement, it was clear that his action had resulted in a considerable delay to construction. On 17 August 1995, the Newbury Weekly News had the headlines, "BYPASS ON HOLD FOR NEW BIDS". Because of the time which had elapsed, it had become necessary to call for new tenders for the construction work. It might have been expected that the tender documents, already scrutinised and commented upon by contractors in 1994, would be almost flawless and certainly sufficiently sound to issue promptly after the restoration of the bypass, but invitations to tender were not sent out until two months later, in September 1995. Six bids were received. The contract was eventually let to Costain Civil Engineering Limited for a tender value of £73.7M. in June 1996. The main contract work commenced in August 1996. Dr Mawhinney's shelving of the bypass had effectively delayed its completion by 1½ years.

Traffic Statistics

The protesters continued to argue against the need for a bypass using traffic statistics. They said that the traffic within the town would rapidly climb back to the pre-bypass levels. This seemed to be based on a misinterpretation of the official traffic forecasts for various roads within the town. The figures had been published to indicate the levels of future traffic both with and without the bypass. No account had been taken of traffic reductions which would result from improvements to local roads because what those improvements would be was unknown. The designs for them could not be put in hand until the final approval for the bypass was given. The sole reason for the production of the figures was to give an indication of the benefits that the bypass would bring. The protesters quoted the rise in traffic, even with the bypass, as evidence that it would be ineffective within a few years as proven by the official figures but no such conclusion can be drawn.

In addition, at Newbury, no such rapid growth of traffic has taken place with its east/west bypass, the M4. As stated in Chapter 1, traffic on the A4 to the west of Newbury and to the east of Thatcham remains manageable even 27 years after the through traffic was removed. The bypass will inevitably induce a percentage of extra traffic (up to 10% according to the Review Team report) and residents of the "rat-run" villages around Newbury will be glad to see it happen, to relieve their plight. The abandonment of the A36 Salisbury bypass will also induce more south coast to Bristol heavy traffic to use the longer but quicker A34 and M4 route.

The Friends of the Earth held a meeting in Newbury on 10 October 1995, advocating a series of measures to solve Newbury's problems including computerised traffic management schemes, priority for buses and high-occupancy vehicles, speed limits and traffic calming, parking restrictions, park and ride schemes, special freight rail services, car-sharing, more school buses and safe cycling and walking routes. Most of these worthy measures are likely to be used in the future, some were already in place, but their main objective appeared to be a reduction in local, not through traffic. There was little support in Newbury for the contention that these proposals provided an alternative to a bypass.

The Friends of the Earth claimed that only five per cent of the traffic at Newbury was long distance through traffic, implying that the bypass would not provide the predicted relief within the town. The various sections of the 7½-mile A34 in and around Newbury carry quite different volumes of local traffic. The Newtown Straight carries only about one quarter of that within the centre of the town. The percentages of through traffic vary from about 75% on the Newtown Straight to about 50% in the town, averaged through the day. The majority of the local traffic occurs during the morning and evening rush hours but at no period could it be said to comprise 95% of traffic on the A34. The reality of the danger of heavy through traffic was highlighted on 29th October 1995 when a 38-ton articulated lorry suffered a brake failure while approaching Newbury town centre on the Sandleford Link and collided with three cars, causing two fatalities.

Many drivers had found "rat-runs" avoiding the A34 through Newbury. This was borne out by the traffic density along the Newtown Straight where a traffic counter had been installed many years previously. The volume of traffic approaching Newbury from the south at that point had begun to level off as regular drivers became weary of the lengthy delays when approaching Newbury and realised that it was preferable to divert down the A343 Andover Road or via Burghclere village and Thatcham or other routes. Even so the increasing need for a bypass was borne out by the traffic recorded by the Tot Hill traffic counter where volumes were now more than 30% higher than the predictions tabled by the DoT at the 1988 inquiry at which traffic forecasts had been provided for high and low economic growth in the UK - see chart. (The National Traffic Forecasts were revised upwards later in 1988 but the actual traffic at Newbury has exceeded even the revised forecasts.)

A34 TRAFFIC AT TOT HILL

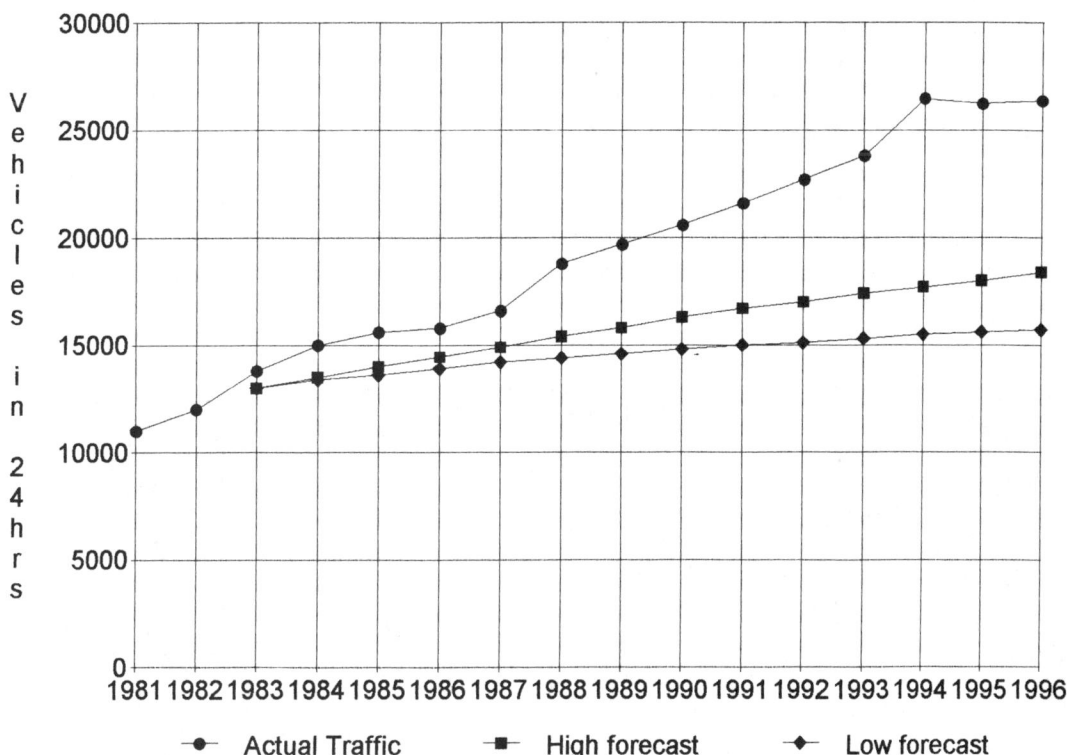

The protesters' claim that through traffic was not the major problem was greeted with some derision in Newbury and was expressed in this sarcastic verse published in the Newbury Weekly News;

The Friends of Earth have told us
That our traffic's mainly local.
They know, 'cos they're from London Town;
I'm just a Berkshire yokel.

I lean on Sainsbury's corner
Watching Britain's biggest queue,
"But only five per cent", they say
"Are really passing through."

So Eddie Stobart's trucks aren't from
Southampton Docks or Poole.
They run a shuttle service
Taking Newbury's kids to school.

Those car transporters spend all day
Between our local dealers,
Collecting cars from Gowrings and
Off-loading them at Wheelers.

Each Friday, right from Beacon Hill
To Chieveley, stuck in rows,
The traffic merely waits its turn
For shopping at Tescos.

Perhaps the moon is made of cheese
And pigs can fly unheeded,
And I'm the Duke of Marlborough
And the bypass isn't needed.

Rallies and Violence

The protesters had turned to mass action. A rally was called at Donnington Castle by Friends of the Earth for July 29th, 1995. About 350 people apparently attended, mainly from outside the Newbury area and were reported to have been transported from as far afield as Leeds, Stratford and Ealing and were dubbed by local residents, "rent-a-mob". An observer who attended the rally reported, *"The people at the rally represented the most smug, self centred and self righteous group I have seen for years. The supporters of the bypass are not stupid or insensitive, but most are practical and caring, for people and the countryside."*

The delay in the letting of the contract gave greater encouragement to the protesters and the numbers occupying camps and tree houses increased in second half of 1995. By the beginning of 1996 there were said to be 7 main camps and 28 others. In December 1995, the Sunday Express carried the headline, "DOLE SENT TO TREE HOUSE", and went on to report that, *"Hippy "eco-warriors", battling to stop a new road are getting dole handouts delivered to their illegal tree-top camps"*. At one camp, *"postmen had to hike half a mile along a muddy canal towpath and avoid several hungry-looking dogs before making their mail drops"*. One postman concerned said that most of the time when he delivered their post, the protesters were still asleep and so he had been leaving the mail in a pile by the camp fire. His supervisor said that he had to deliver to some sort of post-box, so the protesters had made a post box out of pallets. *"As far as I am concerned that is a post-box and deliveries can be made there"*, he said. The news caused anger within Newbury and elsewhere. Letters to the Daily Telegraph were scathing in their condemnation of the protesters, a typical remark being, *"No other country on earth allows itself to be tormented by a Luddite band of protesters having a jolly time while being paid by the taxpayer"*.

"AAAAAAAAAEEEIIIEAAAAAGH! Your dole cheques are here!"

The publicity was picked up and echoed overseas. On 7 February 1996, the New York Times International carried an article and photograph on the Newbury protesters. The Western Australian noted, *"Prising protesters out of the way has become part of road building in Britain"*. Both Dutch and French television carried extensive reports on the activities of the protesters.

Court actions to evict the protesters started in January 1996. On 27 February, appeals were dismissed in the High Court and evictions started in a blaze of publicity on 29 February led by the Under-Sheriff for Berkshire. At times there were more than 40 journalists on the site. It was essential that tree felling should be completed by the beginning of the bird nesting season. Hydraulically operated access platforms, so called 'cherry-pickers', were employed to reach protesters in the trees, the tallest tree requiring the use of a 'cherry-picker' specially imported from Belgium for the purpose. Professional climbers were employed to bring protesters safely to the ground.

A Friends of the Earth national rally in Newbury on 11 February 1996 was followed the next day by mob violence. The Newbury offices of Tarmac were wrecked by 60 demonstrators, some masked. The protesters then invaded the offices of a company recruiting security guards before pushing their way into the Newbury District Council Offices. On the construction site, extremists among the protesters resorted to spiking and wire wrapping of trees causing a major hazard to chain saw operators. A 'caltrop' spike device for damaging horses' hooves was discovered by mounted police. A protester's tunnel, dug to hinder machinery, was occupied in a swift dawn action by the under-sheriff. Over 800 arrests were made and a high proportion of prosecutions resulted in convictions. A typical case involved a female protester who was arrested for throwing urine at a sheriff's officer. She later pleaded guilty to common assault and criminal damage. Her defending solicitor was reported as saying, *"She has no means and relies on money from the Friends of the Earth campaign"*. She was given a conditional discharge. The Friends of the Earth denied paying protesters to be present on the site. The Highways Agency and the main contractors, Costains, were granted injunctions against several protesters preventing them from trespassing on the site, interfering with vehicles or disrupting the work.

Not all contacts between the protesters and officials were negative. A tree known as 'Middle Oak' formed the centre of one of the last camps. Since the DoT wished to retain it as a landscape feature, the protesters agreed to leave their camp peacefully.

The Friends of the Earth placed a full-page advertisement in The Times newspaper on 6 March 1996 condemning the bypass and claiming that the bypass would make little difference to journey times and its benefits would be short-lived. A denunciation of the advertisement was made to the Advertising Standards Authority who upheld the complaint. A Friends of the Earth director was quoted as saying, *"We're fed up with the ASA's bumbling, amateurish treatment of complex environmental issues . . . They should stick to examining claims about cat food and soap powders"*. Told of this comment, an ASA spokesman replied that, *"It is not a complex environmental issue. The issue is whether FoE could provide sufficient evidence to support the claims they have made and the answer was that they could not. It is as simple as that. It is ironic that they should say that, because they make use of our complaints process themselves when they see an advert they don't like."*

The vandalism continued. In August 1996, protesters broke into the Surrey offices of Pinkertons, the company who had been awarded the contract for providing security to the site. A company spokesman was reported as saying that, *"A female member of our staff was attacked and bitten."* In the same month, David Rendel M.P. was sent the cut-off edge of a razor blade in the post with a note accusing him of being an, "eco-rapist". The premises of a company supplying machinery for the bypass, HE Services Ltd at Membury Airfield, was raided. Vehicle brake lines were cut and sand poured into engines. Numerous other instances of damage were reported. The Highways Agency said in October 1966 that £43,000 worth of damage had been caused to security fencing. A house on the Andover Road adjacent to the bypass, which had been purchased by the DoT, was half demolished in a pointless act of vandalism.

The source of funding for the protesters' legal costs and for transporting their supporters to Newbury has never been disclosed. A common view in Newbury was that certain wealthy, well meaning but misguided individuals were making anonymous donations to the protesters' cause.

Miniature snail ©PB

Ecological Protection Measures

The rare miniature snail, vertigo moulinsiana or Desmoulin's Whorl, was discovered on the route in the Lambourn Valley and in June 1996, a coalition of environmental groups applied in the High Court for leave to appeal for a judicial review of the government's handling of the snail issue. The judge said that the application had to fail because the government could not be shown to have acted improperly or irrationally. In practice, and under guidance from English Nature, new and larger snail habitats of 3,200 sq. metres were formed in the Lambourn and Kennet Valleys by carefully moving earth containing reeds and sedges to new locations which have the correct depth of water table.

The Highways Agency were going to considerable lengths to protect the environment, giving the protesters little opportunity for genuine criticism. The measures taken to protect badgers, dormice and bats have already been mentioned. During clearance work, adders and grass snakes were discovered hibernating in a bank in the Tot Hill area. The rabbit warren habitat favoured by such snakes was created artificially close by using breeze blocks and plastic pipes. Because the Environment Agency's measures to encourage otters to migrate back into the Kennet and Lambourn Valleys has proved successful, otter tunnels have been provided where the bypass crosses watercourses.

The Highways Agency commenced the publication and distribution of a Newsletter giving details of the progress of the work and, somewhat belatedly for supporters, stating some of the advantages of the bypass. The Agency pointed out that around 20,000 vehicles a day will be diverted from Newbury's streets, including up to 400 heavy lorries an hour. Over its first 30 years, the bypass was estimated to save 28 lives and avoid 1,800 casualties.

A £250,000 contract was awarded to York Archaeological Trust to undertake the excavation of a Mesolithic site in the Lambourn Valley and was completed in October 1996. The general policy of the Highways Agency is that in-situ preservation is preferable to excavation since the science of archaeology is developing rapidly and future techniques will almost certainly be able to extract more information than present ones.

Other measures to protect the environment make the Newbury Bypass probably the most environmentally friendly road scheme in the UK. The protesters frequently pointed out that 10,000 trees needed to be felled as part of the site clearance work, but ten times as many were to be planted to screen the finished road. A running surface of porous asphalt along the whole length of the road absorbs sound and reduces spray - not without the disadvantages of a higher degree of treatment in frosty conditions and occasional removal of accumulated detritus from the pores. All surface water drainage along the length of the bypass is first directed to oil interceptors. In the event of a major spillage following an accident, a penstock is provided to seal the interceptor and retain the pollutant. For two-thirds of the bypass, the storm water then flows through silt traps and large balancing ponds which prevent surges affecting the local watercourses, into which the ponds eventually drain. These ponds are planted with reeds to create wetland habitats and absorb pollutants such as heavy metals. The remaining one third of the bypass is built on chalk which is an good medium for soakaway drains.

In constructing the bypass, use has been made of crushed concrete from the old Greenham Common runway for building temporary construction roads and tracks. Much of the excavated material on the site has been suitable for sieving, grading and using for such purposes as backfill to bridges or drains and as a 'capping' layer between the earth formation and the road base itself. Instead of being taken off site, surplus excavated material has also been used to provide higher banks for acoustic and visual screening. Cables and a sewer are carried within the steel box girders of the A4 bridge at Speen. Lay-bys are provided at intervals along the bypass. Planning approval has been given for a service area at the southern end of the bypass with access from the western roundabout at Tot Hill.

Protesters at the home of Sir George Young ©PB

Further Violence

None of the environmental protection measures reduced the protesters' determination to use the Newbury Bypass as its focus for anti-road demonstrations. On 8 January, 1997 protesters went to the home of the Secretary of State for Transport, Sir George Young, near Maidenhead and 10 arrests were made for conspiracy to cause criminal damage. On 11 January, the Newbury Friends of the Earth branch called another on-site demonstration with speakers from the London office. Tony Benn also spoke at the rally of the *"brilliant campaign against the Newbury Bypass"* which would *"re-establish civil liberties and democracy which are now being systematically eroded by the criminalisation of dissent through repressive legislation"*. It had become clear that such on-site rallies would almost inevitably end in violence. It was rumoured that travel assistance was provided by supporters' organisations and there is no doubt that many of those attending saw themselves as taking part in a type of environmental pilgrimage. Encouraged by inflammatory speeches, the euphoria of such an event can escalate through competitive high spirits into actual vandalism. Once again the demonstration was followed by violence which caused over £200,000 worth of damage to machinery and the burning down of site offices. Four security guards suffered concussion.

Arson attack on a site office by protesters in January 1997 ©PB

One of those attending, a Nonconformist Moderator, the Rev. John Miller of Reading, was reported as saying, *"The protesters were justified in breaking into the compound and destroying the equipment as they did"*. At a meeting in Newbury on 20 January, Arthur Scargill also expressed his support for the protesters by saying, *"The law has not been handed down in tablets of stone: it is man-made. When your conscience dictates, you have a duty to defy it"*. His speech prompted a response from the area commander of Newbury police; *"I did listen to his encouragement for civil disobedience, his deploring of police behaviour and his wild accusations that the damage and disorder was an MI5 plot. If all these issues were not so serious, I would have thought I was listening to a candidate from the Monster Raving Loony Party"*. The Friends of the Earth attempted to distance themselves from the violence but within Newbury there was general condemnation of the rally's organisers and the prominent speakers supporting them, summed up in a letter to the Newbury Weekly News, *"Their repudiation of the activities of the eco-louts sounds as hollow as a Sinn Fein disclaimer for an IRA bomb"*.

50

One protester, using, among other aliases, that of 'Swampy', briefly achieved almost folk hero status in the media when he delayed work at the site of the Honiton Bypass by remaining in a tunnel dug by protesters. The Reading Magistrate who fined him £500 for damaging a £15,000 theodolite at Newbury took a different view, *"Stop living off the back of society. The court in no way condones your behaviour. Perhaps you should devote your energies into getting work."* The building of the Honiton Bypass and the extension to a runway at Manchester Airport early in 1997 gave the protesters other sites at which to demonstrate and most of them moved away from Newbury, which was clearly becoming a lost cause. A very few remained to provide token acts of protest. (Newbury is used to such persistence among protesters. At Greenham Common, some women protesters remain encamped, years after the airfield itself closed and reverted to civilian uses.) The construction work at Newbury was proceeding to programme in spite of the demonstrations and violence, with an estimated date for completion of autumn, 1998.

The Attitude of the Civil Engineer

Many of those involved with the design and construction of schemes such as the Newbury Bypass have an ambivalent attitude to their work. While they strive to ensure that the works are constructed efficiently and safely, they are also mindful of their responsibility to protect the environment, enhancing it where possible. The Institution of Civil Engineers' 1828 Royal Charter, includes its members' functions as carrying out projects *"for the use and convenience of man"*. Its Rules for Professional Conduct charge civil engineers with having *"full regard to the public interest"*. The investigation into the Newbury Bypass has been so thorough and has involved the public to such an extent that none of those associated with it should suffer any twinge of conscience on environmental grounds. The work has been notable for its fresh approach to contractor/consultant/client relationships using more informal arrangements and a "co-operative management initiative", involving improved dispute resolution procedures. Costain Civil Engineering, Mott McDonald and the Highways Agency are to be congratulated on instituting this policy. Civil engineers as a whole receive scant recognition for their role in society. While the Newbury Bypass is not exceptional in its scale or innovation, it has been designed and constructed to the highest standards and should be regarded as a source of pride to those taking part in it.

The A34/M4 Chieveley Interchange

In the meantime the Highways Agency had run into trouble with the separate but related £33 million scheme to provide an improved and complicated A34/M4 junction 13 at the northern end of the Newbury Bypass near Chieveley. It had originally been hoped that the new interchange could be constructed simultaneously with, or immediately following, the bypass, but in February 1997 the High Court quashed orders for the construction of the slip roads, leading to a re-examination of the design with the possibility of a new public inquiry. In March 1997 the Minister for Roads said that new junction remained in the main road building programme and would go ahead when funds were available. In June 1997, the new Labour Transport Minister, Dr Gavin Strang, confirmed that work on the revamped £33million junction would be delayed until well after the opening of the bypass.

bury Weekly Ne

THURSDAY, JUNE 5, 1997

{AM Theale Tadley Woolhampton KINGSCLERE Woolton Hill Burghclere Highclere HUNGERFORD Kintbury Inkpen Great Bedwyn

Minister puts the brake on £33m. junction

CHAPTER 10

CONCLUSIONS

The average time to build a road scheme in the UK, from public consultation to the opening to traffic, has been quoted as 13 years, clearly indicating a failure of the procedures currently being used. The Newbury Bypass, contrary to the impression conveyed by the protesters, was a straightforward scheme requiring only the normal primary and secondary inquiries. The DoT's chosen route was judged by the independent Inspector to be the best solution on environmental, traffic and economic criteria and had by far the greatest public support. Even so, the time span from public consultation to opening will have been 16 years, of which only 1½ years delay can be attributed to the personal intervention of the Secretary of State for Transport, Dr Mawhinney. If the consultant's initial investigation period is included, the total time will have been 19 years. Many of the later problems arose from the long timescale of the processes. Passions had time to build up. Long periods of silence from government quarters implied that officials and politicians were having second thoughts and gave encouragement to those opposed to the scheme who then financed and put forward new alternatives. Restoration of the scheme after such silences then only inflamed feelings even more.

The sound reasons for the two public inquiry Inspectors' firm recommendations for the western route began to fade in the public memory. Even a former Transport Minister, Steven Norris, in a BBC Panorama TV programme in April 1997, gave it as his opinion that the wrong route was chosen for the bypass, saying that the reason for rejecting the central route was difficulty of construction. He had presumably forgotten the many other overwhelming reasons for the Inspector's strong recommendations against any central route. The protesters seized on Mr Norris's comments as proof that a mistake had been made. In the minds of the public, there developed a "no smoke without fire" suspicion that the protesters must have a valid point.

Delay has a dramatic effect on costs. Apart from increases due to inflation, the additional costs of removing protesters at Newbury escalated enormously. The construction cost of the scheme was quoted as around £74 million; the Highways Agency's estimate for security and policing up to the opening of the bypass was given as £35 million.

Of great importance is that a long timescale drastically affects the future planning by Local Authorities of town development, whether of domestic and commercial buildings, local road schemes, or other projects. Newbury District Council and Berkshire County Council were badly affected in this regard, not knowing when or whether the bypass was to be built and if it was, where it would be routed. Private individuals suffered in a similar way, not knowing whether and to what extent their homes would be affected. Allowing this effect on people's lives to persist for more than a decade must be regarded as the major failure of the present procedures. All parties are agreed that some improvement to the present assessment procedures has to be made. The most essential step forward is to improve drastically the total time taken.

None of the bodies associated with the Newbury Bypass emerge free of criticism of their conduct over the Newbury Bypass. Almost all could have done better. Lessons can be learned by reviewing the conduct of those who played a prominent part either in the inquiries or in subsequent events. Their roles are summarised here, not so much with the intention of allocating blame, but in order to indicate the logical steps for improving future methods of assessing such public schemes.

The Politicians

At several stages of the assessment at Newbury, delays were being attributed to a government reluctance to spend money. There can be little doubt that, on publicly funded road schemes, the government has an incentive to surrender to a green lobby if it appears to be gaining appreciable public acceptance; it provides an excuse for deferring expenditure. It would be unrealistic to expect any government of the day to relinquish its power of approval over schemes at any stage. It will wish to retain control either to limit spending or for some other political reason. However the repercussions of such behaviour can be so serious that, some curbing of the existing autarchy would seem to be necessary. Those wielding such power should be more accountable for their decisions. This can perhaps best be achieved by a method which throws into greater public prominence attempts to delay or cancel the democratic processes.

The DoT, the DoE and the Highways Agency

No information is made available as to how much assessment time is taken up by civil servants or agency personnel and how much is attributable to the deliberations of Ministers; the dates on which briefs are handed to Ministers are not known. However the time spans at Newbury - such as taking four years between public consultation and the publication of Orders - cannot possibly be justified. The role of the civil service administrators in the lengthy drafting of decision letters or the issue of contract documents is particularly suspect. The standard reply that, *"we must get it right"* is not a credible excuse for long-windedness. In industry, documents of equal complexity and importance are expected to be produced in weeks, not months.

The public relations function of the government departments is notoriously weak. The misleading claims of the protesters were allowed to go unchallenged by the Highways Agency before Dr Mawhinney's shelving of the scheme. During the subsequent Review Team's exercise, the Agency said that their publicity had to be suppressed so as to avoid accusations of bias. The Agency's Newsletter was not started until construction work was imminent; too late to be effective. Rather than pay the enormous costs of site security, it would without doubt have been cheaper to finance an early, comprehensive publicity campaign discrediting the wilder claims of the protesters. Such a policy would have reduced the effectiveness of their misleading propaganda which attracted financial support for their activities.

The Protesters and Environmental Organisations

The protesters and their organisations, together with their financial sponsors, carry a heavy responsibility for the disputes and additional costs at Newbury. The Friends of the Earth did not take an active part in the public inquiries. The accusation has been made that public inquiries are somehow rigged and that the outcome is predictable and therefore participating in them is a waste of time. That imputation cannot be justified so far as the Newbury inquiries were concerned. Every witness had a sympathetic hearing and the Inspector and Secretaries of State were influenced by evidence proven to be truthful. The suspicion exists that the protesters' organisations knew that their flawed arguments would be revealed under cross examination and that releasing them to an uncritical media at a later date would have greater impact.

At the very least, it was naïve of their leaders to imagine that the organising of mass on-site rallies would not lead to vandalism by extremists. Some individual protesters were self-confessed anarchists. The considerable financial resources clearly possessed by some of the organisations enabled them to pursue legal actions which appeared to have little chance of success. While few people would deny their right to protest peacefully, their legal failures, their perpetual carping and the violent actions of those perceived to be their supporters merely antagonise public opinion and do a disservice to the greater environmental cause. Their members no doubt consider that they perform a vital role in identifying and publicising instances of environmental damage, but their credibility and usefulness would be enhanced in the eyes of the general public if they also gave positive help and encouragement to such schemes as the extensive environmental protection measures being carried out at Newbury and elsewhere.

Media

The prime objective of newspapers, television and radio is to achieve the maximum numbers of sales, viewers or listeners, respectively. The public at large is attracted by sensational stories. The national media will therefore prefer to report on the protesters' more exaggerated claims or their violent actions rather than the supporters' less newsworthy reasoned arguments. In most articles on the Newbury Bypass, far more space or time was allocated to protesters than supporters. Although there has been public outrage against the vandalism, the inevitable result has been a widespread public belief that the Newbury Bypass causes considerable environmental damage and that a better solution could have been found. By this discrediting of the assessment process, public sympathy has been aroused which tends to fund the protest movement and incur enormous additional public expenditure, almost certainly leading to delay of the bypass and the consequently high local costs to the councils and public around Newbury.

Supporters

The supporters of the bypass do not emerge untarnished. While the protesters were mounting their campaign against the bypass in 1994, by the publication of alternative routes and other criticisms, in the main the local councils and residents adopted a laissez faire attitude and assumed that the government procedures would be sufficient to maintain the momentum of the democratic process. The possibility of politically motivated action was ignored. The Bypass Supporters' Association, representing local residents, provided a lone voice in attempting unsuccessfully to refute the protesters' distorted arguments. Its resources could not compare with the national financial backing available to the protesters. It was not until the bypass was shelved that the local community woke up to its responsibilities and formed what became a successful lobby group. Had the Bypass Forum been convened earlier, it is almost certain that the eighteen month delay would not have occurred.

RECOMMENDATIONS

Since the 1988 Newbury Bypass Public Inquiry, there have been changes to the Highways (Inquiry Procedure) Rules. They are laid down in Statutory Instrument 1994/3263 and came into force on 10 January 1995. They deal with the setting up of the inquiry and lay down time scales for the serving of notices but they do not place any onus on Ministers on when to publish reports or reach decisions.

The primary objective should be the establishment of a speedier, less confrontational, more open and more respected method of assessment in which all those with a genuine interest would be allowed to participate at an early stage and who would therefore feel that they had been given adequate opportunity to express their point of view.

An essential element must be the discovery of the facts in the case. It is difficult to envisage how the truth can be disentangled from exaggeration and falsehood other than by calling for evidence to be presented by witnesses who can be cross-examined in front of an independent Inspector, as in existing inquiries.

Following publication of the scheme concept, which would include indications of the environmental effects, the whole process should begin by calling a one or two-day conference of interested parties under the chairmanship of a Lord Chancellor's Inspector or other unbiased arbitrator to establish a timetable or schedule of events - a Formal Programme - leading up to the final decision point in the scheme and taking into account the complications of the case, the degree of opposition, the public interest and other factors. (This would constitute an extension and strengthening of the 1994 changes to the Inquiry Procedure Rules, which already formalise the pre-inquiry meeting and call for the Inspector to produce a timetable.) This Formal Programme should be well publicised and given a high profile and should form the framework to the democratic process, to be amended only by published justifications. Together with the publication of the Formal Programme, the Inspector would issue a brief report giving a summary of the issues and likely alternatives.

The Formal Programme would alert all parties, including civil servants and politicians, to the probable timetable for their participation so that financial resources and professional staff time can be allocated for a particular period in the future. Only a limited slippage for exceptional circumstances should be allowed. The Formal Programme would be binding on all parties with appropriate compensation paid for non-compliance or, in the case of evidence, disqualification from participation for lateness so as to preclude delaying tactics by the anti-road lobby.

The setting of the Formal Programme would be followed by an intense period of full time daily consultation and meetings, conducted by a Highways Agency project team set up for the purpose and involving local authorities, environmental organisations, residents associations, landowners likely to be affected and any other relevant bodies or individuals. Participants should be required to disclose whom they represent and the source of their funding, so as to prevent undue influence being exerted by covert vested interests.

It is frequently argued that the public inquiry system is unfair since government departments and large organisations can afford to employ lawyers and experts whereas minority groups, who may have an important point of view, do not possess the resources to have their case effectively presented. Free legal aid is not available for public inquiries. (Though successful objectors may recover their expenses.) However, the public inquiry Inspector has appreciably more discretion than a judge in deciding how the proceedings should be arranged and who should be allowed to cross-examine witnesses. The proceedings are generally more relaxed and less intimidating to participants than those in a court of law. If legal and expert aid were made available, it would be virtually impossible at the start to determine the proportion of aid to be allocated to each of the factions requesting it. It is not until the facts have been established that genuine evidence can be separated from distorted data or mischievous falsehood. Even the initial conference and subsequent open discussions would not necessarily eliminate those organisations or groups whose primary objective is to prolong and escalate the cost of the investigation to a point where the project might be abandoned.

At the deadline fixed in the Programme, a proposal would be published which, in the case of a road scheme, may be a bypass route or some other solution. The Environmental Impact Statement (EIS) would be included in the published proposal. Where close alternatives exist, more than one EIS may be needed. Given the public interest in the environment, a public inquiry is nowadays almost inevitable, if only to establish the facts of the case, and this would be included in the programme, followed shortly by the publication of the Inspector's report. All alternatives should be considered at the inquiry and not just the Departments' favoured solution. A period of time would be allowed for comments on the report before the decision is announced. The normal secondary inquiry will probably be required and a report and decision published. The contract documents should then be drawn up, the design engineers and lawyers concerned having been given adequate notice in the programme that the documents would be required within the period allocated. The additional time allowed for tender and construction periods would follow current practice since there is little scope for reduction, particularly since EEC-wide tendering is now required.

The maximum publicity, advertising and continuous reporting should be used at every stage to carry public opinion along with the steps being taken and to give the Formal Programme a respected status. By this means an onus would be placed on politicians so that any intervention by them to the Formal Programme would be used in exceptional circumstances only and be accompanied by a formal statement of convincing reasons. The likelihood of delays to suit political convenience, such as waiting for the end of a parliamentary term to make announcements, would be reduced, the alternative being to risk public opprobrium and parliamentary challenge. The timetable would alert the Treasury well in advance of the funding needed, although a higher proportion of future schemes are likely to be built using private finance.

A major change to the Formal Programme could occur if, for example, outright rejection of all of the alternative solutions occurred. A new process might then need to be initiated, but such a possibility should not be a reason for failing to embark on a set timetable at the outset.

In the case of a project on the scale of the Newbury Bypass, such a system might result in the following timetable;

	Up to Month
Publication of possible solutions	0
Initial Conference followed by setting of Formal Programme	4
Consultation and round of talks with major participants	10
Publication of Preferred Solution and EIS	13
Start of primary Public Inquiry	16
End of Inquiry	20
Report published (before decision taken)	23
Decision taken	26
Secondary Public Inquiry	29-30
Report and decision	32
Contract out to tender	34
Contract signed	37
Work commenced	38
Road opened	64

TOTAL TIME TAKEN	5 Years 4 months

Those who have been involved with past approval procedures will regard these periods as very optimistic, but even this timetable is appreciably longer than similar schemes in continental Europe appear to take, perhaps due in part to the protest movement being less developed there than it is in Britain.

Activities associated with the scheme could overlap. For example, the Environmental Impact Statement could be formulated as the consultation period proceeds; the Inspector could be provided with adequate assistance to compile much of his report during the course of the inquiry; the civil engineering design work could commence with the publication of the preferred solution even though there would be some risk of its being modified or aborted. Tenders could be drawn up during the course of the secondary inquiry. The serving of notices could be made in anticipation of events taking place. Waiting for one activity to finish completely before commencing another defeats the object of a process when overall time is of the essence. Present accountability rules for public servants may require amendment to allow them to initiate tasks which might subsequently prove abortive.

Having pursued an open policy, although peaceful demonstrations must always remain a democratic right, any subsequent trespass and vandalism should be handled with "zero tolerance", the Criminal Justice Act being tightened where necessary. This is what the vast weight of public opinion would support. For most people the danger of anarchy is a greater threat to democracy than a tendency to oppression of extremists.

The influence of the media has to be curbed. Restraints should be imposed similar to those applying during court cases or at times of General Elections. For example, during the public inquiries, proportional television and radio time should be given to the major points of view, contempt proceedings being used when infringements occur.

A not uncommon criticism of bypass inquiries is that they address only one bottleneck whereas the complete strategic route involved should first be open to public investigation. It is the view of all political parties that the UK road network is virtually complete, so that this argument is not likely to recur. However, if it does arise, there is no reason why a primary Formal Programme should not be initiated to cover the whole route, with local schemes following with their own detailed Formal Programmes in the way described.

Many will regard these recommendations as far too radical to gain acceptance among civil servants and politicians but it is certain that a review of current procedures is urgently needed. It is hoped that this history will contribute towards the long-overdue debate on the issue.